Simone Schreijäg

Microstructure and Mechanical Behavior of Deep Drawing DC04 Steel at Different Length Scales

**Schriftenreihe
des Instituts für Angewandte Materialien**

Band 14

Karlsruher Institut für Technologie (KIT)
Institut für Angewandte Materialien (IAM)

Eine Übersicht über alle bisher in dieser Schriftenreihe erschienenen Bände
finden Sie am Ende des Buches.

Microstructure and Mechanical Behavior of Deep Drawing DC04 Steel at Different Length Scales

by
Simone Schreijäg

Dissertation, Karlsruher Institut für Technologie (KIT)
Fakultät für Maschinenbau, 2012
Tag der mündlichen Prüfung: 05. September 2012

Impressum

Karlsruher Institut für Technologie (KIT)
KIT Scientific Publishing
Straße am Forum 2
D-76131 Karlsruhe
www.ksp.kit.edu

KIT – Universität des Landes Baden-Württemberg und
nationales Forschungszentrum in der Helmholtz-Gemeinschaft

KIT Scientific Publishing 2013
Print on Demand

ISSN 2192-9963
ISBN 978-3-86644-967-1

Microstructure and Mechanical Behavior of Deep Drawing DC04 Steel at Different Length Scales

Zur Erlangung des akademischen Grades

Doktor der Ingenieurwissenschaften

der Fakultät für Maschinenbau

Karlsruher Institut für Technologie (KIT)

genehmigte

Dissertation

von

Dipl. Ing. Simone Schreijäg

aus Stuttgart

Tag der mündlichen Prüfung: 05. September 2012

Hauptreferent: Prof. Dr. rer. nat. Oliver Kraft

Korreferent: Prof. Dr. rer. nat. Peter Gumbsch

Acknowledgements

Die vorliegende Arbeit entstand während meiner Zeit als Doktorand am Karlsruher Institut für Technologie (KIT) am Institut für angewandte Materialien (IAM). Im Folgenden möchte ich mich bei den Personen bedanken, die zum Gelingen dieser Arbeit beigetragen haben.

Herrn Prof. Dr. Oliver Kraft danke ich für die Aufnahme an seinem Institut und für die Betreuung dieser Arbeit. Außerdem bedanke ich mich für sein Interesses an dieser Arbeit und die fachlichen Diskussionen, die zum Gelingen dieser Arbeite beigetragen haben.

Herrn Prof. Dr. Peter Gumbsch danke ich für die Übernahme des Korreferats.

Meinem Betreuer Dr. Reiner Mönig danke ich für die freundschaftliche Zusammenarbeit. Seine Ideen, seine Hilfe im Labor und unsere langen Diskussionen haben im Wesentlichen zum Gelingen dieser Arbeit beigetragen. Vielen Dank!

Meinen Kollegen im Graduiertenkolleg danke ich für die gute Zusammenarbeit und das freundliche Arbeitsklima.

Meinen Kollegen am IAM-WBM danke ich für die freundschaftliche Zusammenarbeit und das gute Arbeitsklima. Ganz besonders bedanke ich mich bei Moritz Wenk, Sofie Burger, Sven Bundschuh, Chen Di, Andreas Sedlmayr und Daniel Kaufmann für Ihre ständige Bereitschaft zur Hilfe, die vielen fachlichen Diskussionen sowie persönlichen Gespräche.

Bei meinen Eltern möchte ich mich von ganzem Herzen für die Unterstützung auf meinem bisherigen Lebensweg bedanken. Meinem Vater danke ich besonders, da er mich letztendlich ermutigt hat, meine Doktorarbeit zu beginnen. Da er selber den Abschluss dieser Arbeit nicht mehr miterleben darf, widme ich ihm diese Arbeit.

Acknoledgements

Meinem Ehemann Moritz danke ich für seine Liebe und Unterstützung bei all meinem Entscheidungen in den letzten 11 Jahren und dass er auch in besonders schweren Zeiten immer für mich da gewesen ist.

Kurzzusammenfassung

Tiefziehstähle werden in vielen Bereichen, wie z.B. in der Automobilindustrie, eingesetzt. Ein sehr wichtiges Kriterium dieser Materialien ist deren mechanisches Verhalten. Grundsätzlich betrachtet sind Stähle Verbundmaterialien deren makroskopisches mechanisches Verhalten vom Verformungsverhalten der einzelnen Strukturelemente, z.B. Körner, Ausscheidungen, Phasen sowie der zugehörigen Grenzflächen, bestimmt wird. In dieser Arbeit wurde ein nichtlegierter Tiefziehstahl DC04 untersucht. Zu Beginn wurde die Mikrostruktur analysiert, anschließend wurden mechanische Experimente an ausgewählten Mikrostrukturelementen durchgeführt und danach massive Proben mechanisch getestet. Diese Experimente wurden an Stahlproben aus der gleichen Charge, nach dem Warmwalzen, Kaltwalzen und Wärmebehandeln, durchgeführt, um den Einfluss unterschiedlicher Mikrostrukturen und unterschiedlicher Verformungsgrade zu untersuchen. Zusätzliche *in situ* Heizversuche im Rasterelektronenmikroskop dienen zur Untersuchung der Entwicklung der Mikrostruktur des kaltgewalzten Materials. Diese Experimente, zusammen mit den Proben aus den unterschiedlichen Prozessschritten, ermöglichten die physikalische Charakterisierung der Mikrostrukturentwicklung. Zwei mögliche Ursachen wurden für die beobachteten Unterschiede in den Mikrostrukturen diskutiert. In einem ersten Gedankenexperiment konnte gezeigt werden, dass die Mikrostrukturentwicklung aus einem Wettbewerb zwischen Kornwachstum und Erholung besteht, was beides durch die Reduktion der plastischen gespeicherten Energie getrieben wird. Eine weitere Möglichkeit wäre, dass bei erhöhten Temperaturen eine erhöhte Anzahl an wachsenden Körner vorhanden ist, die zu einem Gefüge aus kleineren Körnern nach dem Wärmebehandlung führen würde. Zusätzlich wurde herausgefunden, dass Subkornbildung und die Nukleation neuer Körner eine untergeordnete Rolle im DC04 Stahl spielen. Mikrodruckversuche wurde an Säulen mit Durchmessern von 0.5 µm bis 22 µm durchgeführt, um das orientierungsabhängige Spannungs-Dehnungs-Verhalten zu untersuchen. Für Probengrößen kleiner als 2 µm folgt DC04 Stahl dem mechanische Verhalten von α-Eisen. Für größere Säulen konnten Abweichungen festgestellt werden, die dem mechanischen Verhalten von massivem DC04 Stahl ähneln. Experimente an Säulen aus Bikristallen und Polykristallen zeigten, dass das mechanische Verhalten des DC04 von Korngrenzen nicht beeinflusst wird. Dies ist in Übereinstimmung mit makroskopischen Zugversuchen, bei denen das mechanische Verhalten von der Probenorientierung relativ zur Walzrichtung abhängt. Mit einem Voigt-Model, das die Texturinformationen mit den orientierungsabhängigen Verformungsdaten kombiniert, war es möglich das anisotrope Verformungsverhalten des Massivmaterials nachzubilden. Diese Ergebnisse deuten an, dass das makroskopische Verhalten dieses Stahls aus der Anisotropie der Textur resultiert.

Abstract

Deep drawing steels are used in many fields such as the automotive industry. A very important criterion for these materials is their mechanical behavior. Fundamentally, steels are composites where the deformation behavior of structural units such as grains, precipitates, phases and their inherent boundaries affect the macroscopic mechanical response of the material. In this work, a simple non-alloyed deep drawing steel was investigated. First, the microstructure was analyzed, then small scaled mechanical experiments on selected microstructural units were performed and then bulk steel samples were mechanically tested. Experiments were performed on steel samples as received from the same batch after hot rolling, cold rolling and a heat treatment step in order to investigate the effect of different microstructures and degrees of deformation on the deformation behavior. Additional *in situ* heating experiments in the scanning electron microscope were performed to investigate the evolution of the microstructure starting from cold rolled steel. These experiments, together with the samples from different production steps, allowed for the physical characterization of the microstructural evolution. Only minor effects of subgrain formation or nucleation of new grains during recrystallization were found to be present. Two different aspects were discussed that explain the differences in the observed microstructures after the different heating temperatures. In a first Gedankenexperiment it was shown that the formation of the new microstructure at the different temperatures is governed by a competition between grain growth and recovery, both driven by the reduction in stored plastic energy. A further possibility may be that the observed differences in grain size are an effect of the increasing number of growing grains with increasing temperature. Microcompression experiments on pillars with diameters ranging from

0.5 μm to 22 μm were used to investigate the microscopic and orientation dependent stress-strain response. For sample with sizes smaller than 2 μm the mechanical response of DC04 follows the behavior of α-iron. For larger samples deviations were identified that result in a mechanical behavior that is similar to bulk DC04. Experiments on pillars from bicrystals and polycrystals showed that the behavior is not strongly influenced by the presence of grain boundaries. This is in agreement with the results found in the macroscopic tensile tests where the response of the material depended on the orientation of the sample relative to the rolling direction. Using a Voigt model that combines the texture information with orientation dependent deformation data, as obtained from pillar tests, it was possible to reproduce the deformation anisotropy of the bulk material. These results indicate that the macroscopic behavior of this steel is purely a result of the anisotropy of the texture.

Contents

1.   Introduction

Deep drawing steels are important materials in the automotive industry due to their applications as vehicle bodies, fenders, floor or door panels. Dependent on their application, they have to provide different mechanical properties as for example high yield strengths, tensile strengths and high fractures strains. During deep drawing, deformation degrees of up to 70 % can locally occur in these materials and therefore excellent formability properties should be provided. A further important aspect is the deformation behavior of the material during crashes. To protect the occupants inside the vehicle, the material is expected to have high strengths and deformability, so that most of the energy that is released by the impact, is consumed during the deformation. All these different demands lead to a high variety of materials that were developed within the last decades. In Figure 1.1 different types of steels are depicted that are generally used in automotive applications, dependent on their fracture strains and strengths. Figure 1.1 shows that the combination of high strengths and high fracture strains is difficult to realize. Important representatives of low carbon deep drawing steels are interstitial free steels (IF) that contain high Ti concentrations, fine grained IF steels and DP steel that have a ferrite/martensite microstructure. Within the last decades a new class of high alloyed steels were developed which contains high concentrations of manganese. These materials exhibit high strengths and high ductilities at the same time due to phase transformations [1-3] or twinning processes [4, 5] that occur locally in the material during deformation. One further advantage is the high amount of energy that can be absorbed during the phase transformation or twinning processes. As a consequence, these materials have great potential for automotive applications.

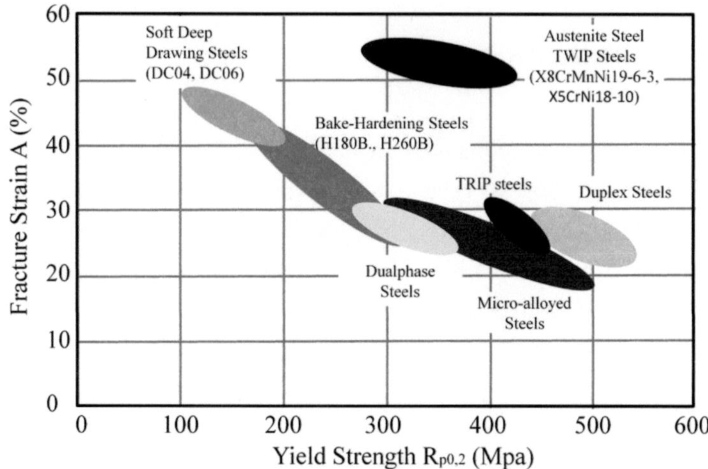

Figure 1.1: Important deep drawing steels that are used for automotive applications [6]

Modern steels are complex materials often containing several alloying elements and different phases, for example ferrite, martensite, bainite or cementite which are not in thermodynamic equilibrium. All phases have very different mechanical properties and therefore the overall mechanical behavior of this composite material is strongly influenced by its microstructure. The spatial distribution of the different phases and the texture of the polycrystalline material is a result of the alloying elements and the processing conditions during steel production.

The experimental optimization of a process chain is very time and cost intensive. Attempts are made to reduce cost by simulating the material response during different process conditions. For high alloyed steels these simulations can be complex and a route has to be established for validation of such models. The Graduiertenschule 1483 (Research Training Group), research area A, focuses on "Process chains of sheet metals" with the overall aim to simulate the whole process chain of a deep drawing steel.

During the production of deep drawing steels, the material is exposed to different process steps such as hot rolling, cold rolling and a final heat treatment (Figure 1.2). All these processes have an impact on the microstructure of the final sheet metal and therefore on its mechanical response. In this research area, phase field models (A3) and texture models (A1) are developed to simulate the evolution of the microstructure and texture during heat treatment. Further projects develop phenomenological models (A4) or micro mechanical models (A2) to simulate the mechanical response of the heat treated sheet metal during deep drawing. The development of these different material models requires a variety of experimental data that give information about the evolution of the microstructure as well as the deformation behavior of the sheet metal after the different process steps. This data is required not only as input data for the models but is also provided for validation and optimization of the simulations.

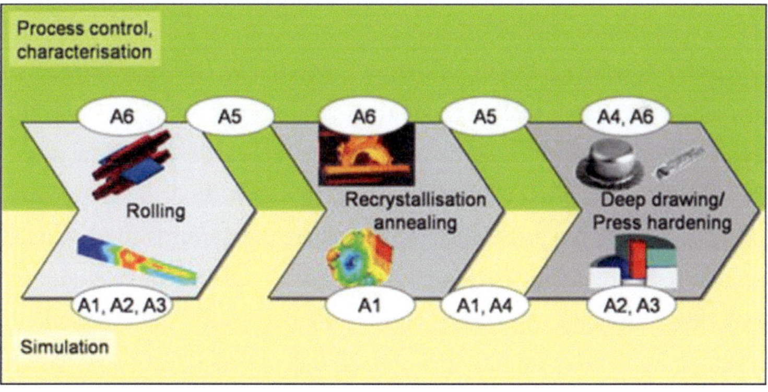

Figure 1.2: Schematic of the process chain that is addressed by the research area A within in the Graduiertenschule 1483

In research area A, the focus is on a non-alloyed low carbon DC04 steel as a starting material for developing an understanding of the physical process and to simulate the process chain. This steel is also the material under investigation of this thesis. It has a relatively simple chemical composition and exhibits a single phase ferritic microstructure. Several aspects of this material have been investigated experimentally in this thesis. In all cases attempts were made to correlate the experimental results with underlying processes with rather simple descriptions. More detailed descriptions and advanced modeling of this material can be found in the Ph.D. theses of the project partners within Graduiertenschule 1483.

In Chapter 4 of this thesis the formation of the recrystallization textures was investigated by heating experiments at different temperatures and heating rates. During these experiments, the local microstructure was monitored by EBSD which is a suitable method to investigate the local microstructures at high resolution. These experiments gave new insights into nucleation processes, grain growth and the evolution of dominant texture components in DC04 steel.

In order to predict the deformation behavior of the final sheet metal, deformation processes that occur on the micro scale need to be understood. An efficient way of probing the mechanical behavior on the micrometer scale is microcompression testing [7]. This method is described in Chapter 5. It gives the possibility to investigate the deformation behavior of samples with different crystal orientations, the influence of grain boundaries, alloying elements and stored plastic deformation. The information obtained on the texture of the material and the data that describe the local deformation behavior were combined in Chapter 6 in a coarse estimation that can roughly predict the macroscopic anisotropic deformation behavior of the sheet metal.

Overall, this thesis gives an overview of the evolution of the local microstructure during heat treatment and the deformation behavior of

DC04 steel on different lengths scales. Besides providing insight into several isolated aspects associated with mechanical size effects, microstructure evolution and texture evolution it attempts to combine microstructural information with small scale mechanical data to describe the macroscopic stress-strain response of a sheet metal. The focus of this work is on experimental research and the models used here are neither accurate nor precise but hopefully will help to understand the physical effects that govern the microstructural evolution as well as the stress-strain response of DC04 sheet metal.

2.   Background

The process chain of deep drawing DC04 steel is composed of different process steps such as hot rolling, cold rolling and a subsequent heat treatment procedure. All these processes have an impact on the microstructure and the deformation behavior of the material. The aim of this work is to investigate the microstructure of the DC04 steel after each process step and the resulting deformation behavior. In this work, the microstructural investigations were performed by Electron Backscatter Diffraction (EBSD). Section 2.1 gives a short introduction on different possibilities to represent microstructural data that were determined by EBSD. Over the last decades, the influences of hot rolling, cold rolling and heat treatment processes on several fcc and bcc metals were investigated by different research groups. Section 2.2.1 and Section 2.2.2 give a short summary on resulting textures and microstructures with a focus on low carbon steels. In Section 2.2.3, some of the basic mechanisms that control recovery and recrystallization processes during heat treatment of strongly deformed material will be presented.

The macroscopic deformation behavior of a metal is related to its microstructure, and may be viewed as a composite material consisting of small elements with sizes smaller than the characteristic length of the sample microstructure. In order to elucidate the macroscopic behavior and its dependence on the microstructure it is useful to investigate the mechanical behavior of such small samples. An efficient method to probe the deformation behavior of single crystalline or polycrystalline samples with diameters of several μm is the microcompression technique. This technique has been widely used to investigate the deformation behavior of several fcc and bcc metals. Some relevant results will be summarized in section 2.3 as well as some mechanisms that may explain the strengthening and the size dependent deformation behavior of micro pillars.

2.1   Description of crystal orientation and texture

The two most common methods that are used to determine crystal orientations are x-ray diffraction and the Electron Backscatter Diffraction (EBSD). The limiting factor that controls the spatial resolution of x-rays is the size of the beam spot which lies in the range of several mm for conventional x-ray diffractometers. Therefore, this method is typically used to determine the texture and other spatially averaged information such as the average grain size. In this work, we were interested in investigating local orientation changes within and between different grains and therefore it was necessary to acquire crystal orientation data with very high spatial resolution. For this reason EBSD was chosen as the adequate technique which has a spatial resolution down to ~20 nm. A detailed description of the EBSD technique, the analysis of the diffraction pattern and of several methods that are used for representing texture and crystal orientations can for example be found in the literature of Schwartz et al. [8] and Engler et al. [9]. The following paragraphs will give a short overview on three different methods that have been used in this work to represent crystal orientations and texture components.

2.1.1   Pole figures

Pole figures describe the orientation of selected crystals relative to the sample coordinate system which also serves as the reference system. The pole figures used in this work are based on a stereographic projection. An example of this projection is shown in Figure 2.1. A sphere is drawn around the center of the unit cell of the crystal whose orientations will be determined. A projection plane is drawn at the great circle parallel to the xy-plane of the sample coordinate system (equator). The directions in the crystal that are of interest (e.g. <111>) are extended until they intersect the surrounding sphere. Lines are drawn from the intersection points to the south pole (projection center) of the sphere. This line intersects the spherical projection plane at a point that is then used to represent the crystal orientation. The pole figure is a plot of the projection plane containing all the measured crystal orientations where each three dimensional crystal direction is plotted as a single point onto the projection plane. By plotting density maps onto this plane, information of the texture can be displayed. Pole figures are very common in x-ray diffraction but can also be generated from EBSD data. The advantage of the stereographic projection is that it conserves the angles between different crystal directions and therefore crystal symmetry is also visible in the pole figure.

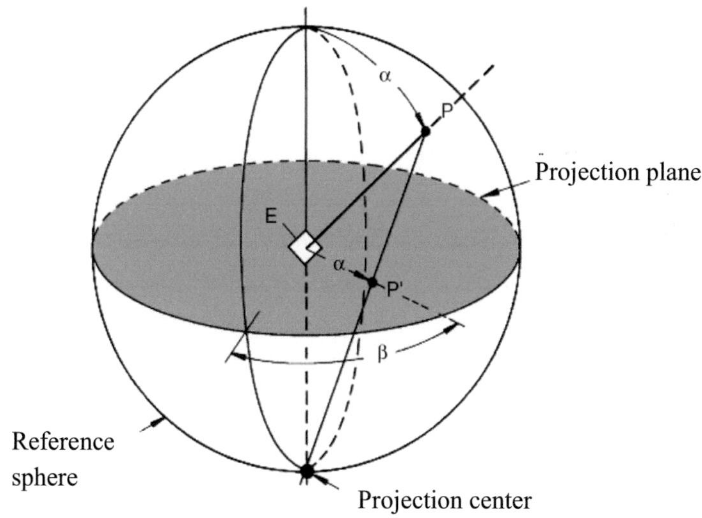

Figure 2.1:Principle of a stereographic projection [10]

Figure 2.2 shows an example of three pole figures that were measured for a DC04 steel by EBSD. In the case of sheet metals that are exposed to rolling processes the axes of the sample coordinate system are usually defined as rolling direction (RD) and transverse direction (TD). In materials that have a cubic crystal structure two pole figures are generally necessary to present the full crystal information. In this work, the {111} , {110} and {100} pole figures are typically plotted together.

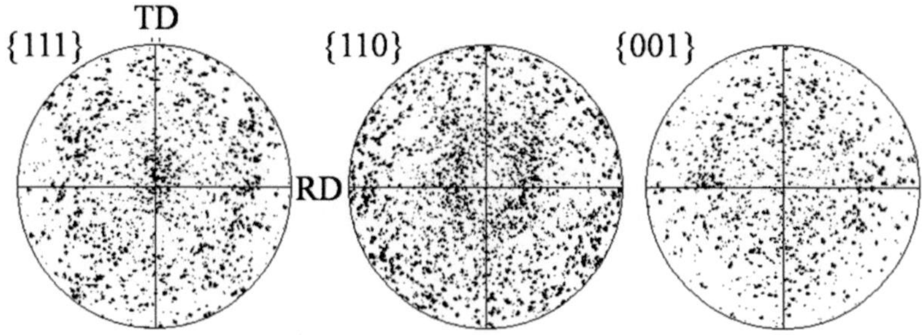

Figure 2.2: {111}, {110}, {100} pole figures of cold rolled DC04 steel

Another way of representing orientations is the use of crystal directions in the form of {uvw}<hkl>. In this case {uvw} is the crystal direction along the sample normal (z direction) and <hkl> the direction along the x direction. In EBSD the z axis is called normal direction (ND), the x axis rolling direction (RD) and the y axis is call tilt direction (TD) or transverse direction.

2.1.2   Inverse pole figures

Inverse pole figures describe the orientation of the crystal relative to a selected sample direction. They display crystal orientations along to a specific axis of the sample coordinate system. According to the cubic crystal system of DC04 steel, the measured crystal orientations are plotted in a standard orientation triangle which is defined by <100>, <110> and <111> orientations. An example of an inverse pole figure is shown in Figure 2.3a. The standard triangle can be highlighted in colors as shown in Figure 2.3c and the investigated EBSD area plotted with the colors that correspond to the crystal orientations (Figure 2.3b). In cubic crystal systems the full three dimensional information about the crystal orientations are given by plotting the inverse pole figure along two directions of the sample coordinate system.

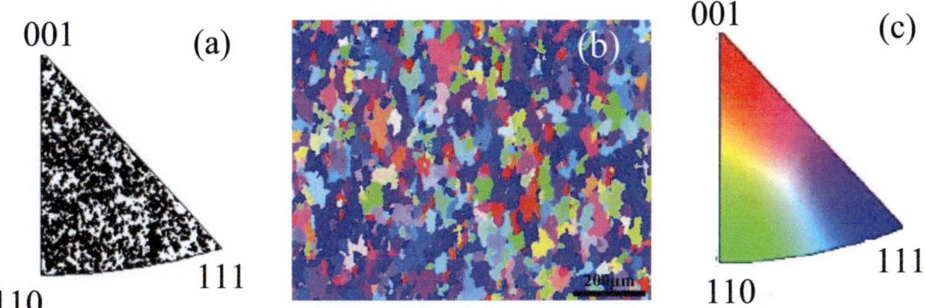

Figure 2.3: (a) Inverse pole figure of a heat treated DC04 sample with cubic crystal structure along the normal direction. (b) is an inverse pole figure map. In this map the data points in (a) are color coded according to (c) and plotted at their position on the sample.

2.1.3 Euler space

Crystal orientations that are determined by EBSD can be expressed in terms of Euler angles for example in the Bunge notation. The Bunge notation [11] describes the rotational transformation of the sample coordinate system, which serves as the reference system, into the crystal coordinate system by sequential rotations around the z-x'-z' axes with $\phi1$, Φ and $\phi2$ which is depicted in Figure 2.4.

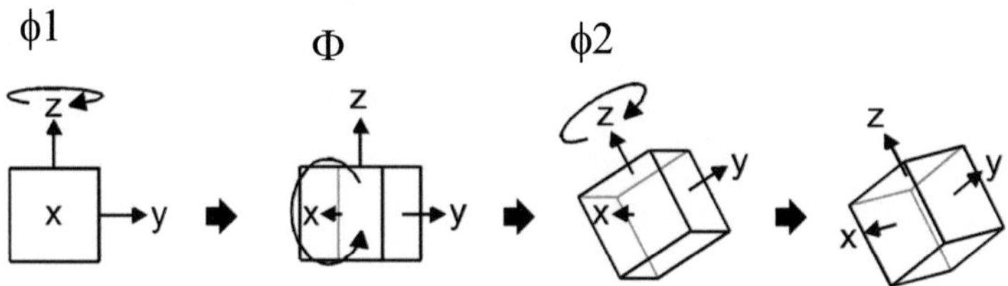

Figure 2.4: Schematic rotation of the sample coordinate system into the crystal orientation by using the Euler angles in the Bunge notation [12]. The sample coordinate system is sequentially rotated around the z-x'-z' axes by using the Euler angles $\phi1$, Φ and $\phi2$.

Every orientation that is expressed in terms of Euler angles can be represented as a point in the Euler space. The Euler space is a 3D coordinate system whose axes are represented by the three Euler angles $\phi1$, Φ and $\phi2$. For materials with a cubic crystal symmetry, Euler angles are ambiguous and equivalent orientations result in a number of different spatial points in the Euler space. As a consequence the Euler space can be reduced to $0<\phi1<\pi/2$, $0<\Phi<\pi/2$ and $0<\phi2<\pi/2$. An example of an Euler space including data for a material with bcc crystal structure is shown in Figure 2.5a. For better visualization, the Euler space can be divided into a series of planar cross sections parallel to one axis with equal distance of for example five degree (Figure 2.5b).

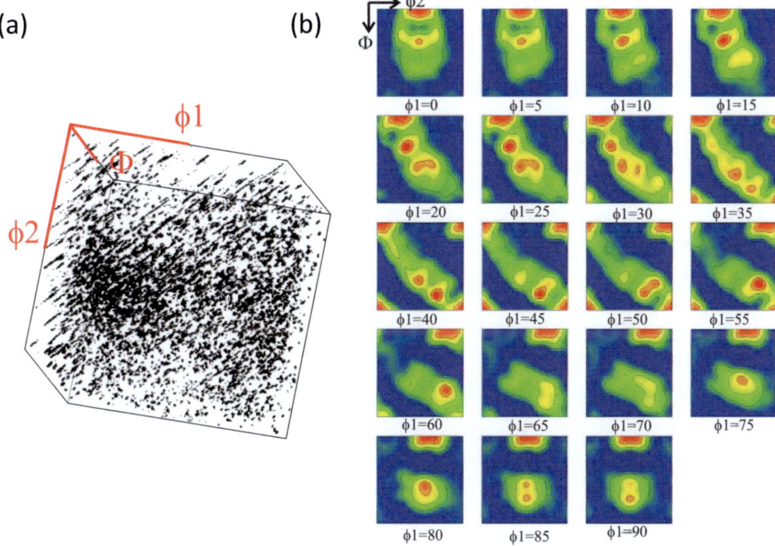

Figure 2.5: (a) Orientation distribution of a DC04 steel plotted in the Euler space and (b) the corresponding serial section through the Euler space parallel to φ1 with the calculated ODF.

2.1.4 *Orientation distribution function (ODF):*

In general, individual poles in a pole figure or Euler space represent a given orientation. However, a quantitative description of these poles is difficult, as all data points in the EBSD map appear as overlapping spots in the Euler space or pole figure if they have same orientations. For a quantitative evaluation of the texture an orientation distribution function (ODF) of these poles needs to be calculated. The ODF is a probability density function $f(g)$ defined by the volume fraction dV/V of crystals that have the orientation g which is expressed in terms of Euler angles $\phi1$, Φ and $\phi2$ [13]

$$f(g) = \frac{dV/V}{dg} \quad , g = \{\phi 1, \Phi, \phi 2\} \tag{2.1}$$

where V is the sample volume and dV is the volume of all crystallites i with the orientation g within and orientation element dg [14]. A detailed description of the calculation of ODFs can be found in the literature [13, 14].

2.1.5 *Density of geometrically necessary dislocations calculated from EBSD data*

In a perfect crystal, the orientation does not change when different locations of the sample are investigated. This is different in real crystals. There the crystal orientations can vary which is accomplished by lattice defects. Common defects that are responsible for the rotation of the crystal lattice are dislocations which can be induced by plastic deformation. Even in annealed metals small changes of the crystal orientations can be observed that are significantly below 1°. During plastic flow, dislocations nucleate and multiply and may lead to misorientations within a crystal. These misorientations correlate with the number of dislocations that are introduced into the lattice. The Euler angles that are measured by EBSD can be used to calculate the misorientations between two data points. Within the last two decades, several methods have been developed to determine the dislocation density from EBSD measurements. In general, only the density of geometrically necessary dislocations (GND) can be determined by the misorientation angle. The statistically stored dislocations do not contribute to orientation changes and are therefore not accessible by EBSD. These dislocations do not lead to orientation changes because the effect on the crystal orientation cancels out. An example for this are two dislocations with opposite Burgers vectors in close proximity which do not lead to a change in orientation as long as distances larger than their spacing are considered. Several concepts exist that allow to calculate the density of GNDs from EBSD data. The first group of concepts is based on the calculation of a dislocation density tensor α_{pi} that was introduced by Nye in 1953 [15]. The components of this tensor can be determined by calculating the rotational gradients $g_{ij,k}$

$$\alpha_{pi} = e_{pkj}\, g_{ij,k} \tag{2.2}$$

where e indicates a permutation symbol. The orientation gradient is then calculated by dividing the minimum misorientation $\Delta\Phi$ between two data points by their distance d in the EBSD scan.

$$g_{ij,k} = \frac{\Phi_{(2)ij} - \Phi_{(1)ij}}{d_k} \tag{2.3}$$

This approach was used by Demir and Raabe [16, 17]. Similar approaches were used by Pantleon [18, 19]. In their studies they used quaternions in order to calculate the curvature of the lattice. For calculating the dislocation density tenor, different assumptions have to be made. In fcc metals 12 glide systems t exist, that can be activated during deformation and each glide system is defined by a Burgers vector b and a glide direction l. The dislocation density tensor is a sum over all individual dislocation densities that are activated by slip.

$$\alpha_{pi} = \sum_{t=1}^{N} b_i^t l_j^t\, \rho^t \tag{2.4}$$

In order to calculate the nine components of the dislocation density tensor, different assumptions have to be made, which of the independent slip systems is active. This leads to some uncertainties in the calculated GND density. A related problem results from the EBSD technique itself. In most studies the EBSD data are only acquired by two-dimensional scans and therefore the gradient information on the third dimension of the sample is lacking. As a consequence the corresponding parts in the dislocation density tensor cannot be determined.

Due to all these limitations, a simpler concept was chosen here. The estimation was performed by considering a cube with an edge length u and realizing that its initially parallel planes are distorted by the presence of dislocations (see Figure 2.6). Placing a single straight edge dislocation inside the cube leads to an elongation of one edge by the Burges vector b. The additional half lattice plane leads to a bending and therefore to a tilt angle between two side faces against each other by $\theta = b/u$ approximating $\tan \theta = \theta$. This concept can easily be extended to more dislocations where n dislocations lead to a tilt angle of $\theta = n\,b/u$. Using the dislocation density as the total line length per volume $\rho = n/u^2$ therefore gives $\theta = \rho \cdot u \cdot b$. With this relation the misorientation θ as determined from EBSD measurements can be interpreted. When the data were recorded on a grid with spacing u, the dislocation density can be estimated by multiplying the misorientation angle θ according to

$$\rho = \theta/(u \cdot b) \tag{2.5}$$

It has to be emphasized that this treatment does only give estimates for the dislocation density and does not account for the different slip systems.

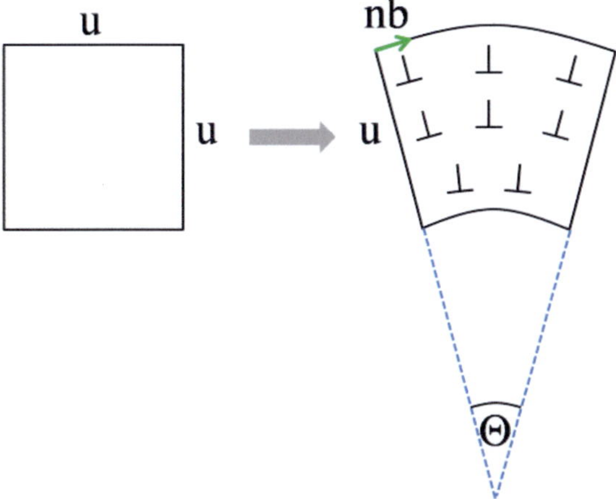

Figure 2.6: Schematic bending of cube with a length u due to the introduction of parallel edge dislocations into the lattice. For this case n=8.

2.2 Processes during steel production and their influence on the microstructure of bcc metals

2.2.1 Hot rolling

Hot rolling processes for low carbon steels are usually carried out in the austenite region (γ-region) [20-24]. The material is usually exposed to several rolling passes in which the microstructure can recrystallize due to the high temperature during deformation. After the last rolling step a weak rolling texture {001} <110> can develop in the center layer of the sheet metal. Raabe et al. [23, 24] investigated that a negligible texture gradient develops trough the sample thickness if the hot rolling procedure takes place in the γ-region. After hot rolling, the subsequent cooling process leads to a phase transformation from the fcc into the bcc crystal structure which leads to a random texture with equiaxed grains. Jonas et al. [25]

observed that the resulting microstructure after the hot rolling process depends on the temperature of the hot rolling process. At higher temperatures, the microstructure can recrystallize during the deformation process which results in relatively large grains after cooling. This recrystallization temperature depends on the alloying elements and the amount of stored energy due to the introduction of dislocations during rolling. For low carbon steels this temperature lies around 900°C. If hot rolling takes place below this temperature, the γ-grains do not recrystallize during hot rolling and the grains remain strongly deformed and elongated along the rolling direction. This leads to a microstructure that consists of smaller grains after the $\gamma \rightarrow \alpha$ transition.

2.2.2 Development of cold rolling textures in bcc metals

The evolution of the microstructure during cold rolling of bcc metals, low alloyed and high alloyed steels became of great interest in the 20th century due to their industrial importance in the automotive industry. The microstructure and texture of the cold rolled steel sheets have a major influence on the deformation behavior of the annealed material. Therefore significant effort is still devoted to research in the fundamental mechanisms that cause the formation of cold rolling textures in these materials. During the last 20 years, deep drawing steels with high contents of alloying elements e.g. Mn or several phases, e.g. ferrite/martensite, became important materials because of their high strengths and high deformability. Therefore, the influence of alloying elements, different phases and precipitates on the evolution of the microstructure during steel production became important. Information about cold rolling textures of more complex steels can be found in the literature [26]. This chapter will focus on the development of cold rolling textures of low alloyed steel with a ferrite microstructure. Raabe, Hölscher, Lücke and Inagaki [22, 23, 26-28] investigated the texture evolution during cold rolling processes of several bcc metals such as Ta, Nb, Mo and W as well as low alloyed steels. These experiments were all performed by x-ray diffraction with a $Mo_{K\alpha}$ radiation.

In general, bcc metals and their alloys tend to form fiber textures during deformation especially during cold rolling processes. Table 2.1 shows fibers that typically develop during cold rolling of bcc metals and their locations in the Euler space. Within these fibers, some relevant components exist that dominate the texture, which are also listed in Table 2.1. Very common fibers that form during rolling processes of low carbon steels are the α- and the γ-fiber. In the α-fiber, the <110> orientations are aligned along the rolling direction of the sheet metal. Orientations perpendicular to this direction e.g. parallel to the sheet normal can be random. In terms of Euler angles (see Figure 2.4) this means that $\phi1=0$, $\Phi=0°..90°$ and $\phi2=45°$. $\phi2=45°$ ensures that the <110> direction is oriented along the rolling direction and $\Phi=0°..90°$ allows all direction that are perpendicular to <110> to be oriented perpendicular to the rolling direction (see Figure 2.4). The γ-fiber consists of <111> orientations that are aligned parallel to the sheet normal whereas the orientations parallel to the RD can be random. In this case, the first Euler angle $\phi1$ can lie between $0°$ and $90°$ and the other Euler angles have fixed values of $\Phi=54.7°$ and $\phi2=45°$. The γ-fibers are often found in polycrystalline thin films [29]. Due to the low surface energy of the <111> surface, grains with this orientation are preferred during growth.

Fiber	Fiber axis	Relevant components belonging to the fiber	Euler angles φ1, Φ, φ2
α-fiber	<011> ‖ RD	{001}<110>, {112}<110>, {111}<110>	0°, 0°. 45° - 0°, 90°, 45°
γ-fiber	<111> ‖ ND	{111}<110>, {111}<112>	0°, 54.7°, 45° - 90°, 54.7°, 45°
η-fiber	<001> ‖ RD	{001}<100>, {011}<100>	0°, 0°, 0° - 0°, 45°, 0°
ζ-fiber	<011> ‖ ND	{011}<100>, {011}<211>, {011}<111>, {011}<011>	0°, 45°, 0° - 90°, 45°, 0°
ε-fiber	<011> ‖ TD	{001}<110>, {112}<111>, {111}<112>, {011}<100>	90°, 0°, 45° - 90°, 90°, 45°

Table 2.1: Important fibers for bcc metals [23], Dominant components and their locations in the Euler space. RD: rolling direction, ND: normal direction, TD: transverse direction

Figure 2.7 shows the reduced Euler space for a cubic crystal whose axes are the Euler angles φ1, Φ and φ2 with the fibers that are listed in Table 2.1 and their relevant components. The {111} <110> orientations form the intersection between the α- and in the γ-fibers in the Euler space.

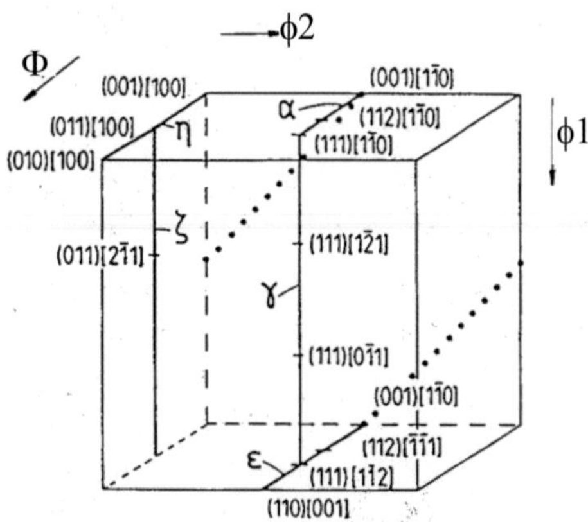

Figure 2.7: Reduced Euler space with fibers that are known to develop in bcc metals [22]

The Euler space in Figure 2.7 shows that the γ-fibers run along φ1 and that the α– and the ε-fibers can be found at φ1=0 and φ1=90°. From this it is evident that the cross sections should be chosen parallel to φ1 and commonly a distance of 5 degree between each cross section is chosen. Figure 2.8 shows the cross sections through the Euler space and the most prominent components that were found for the bcc metals.

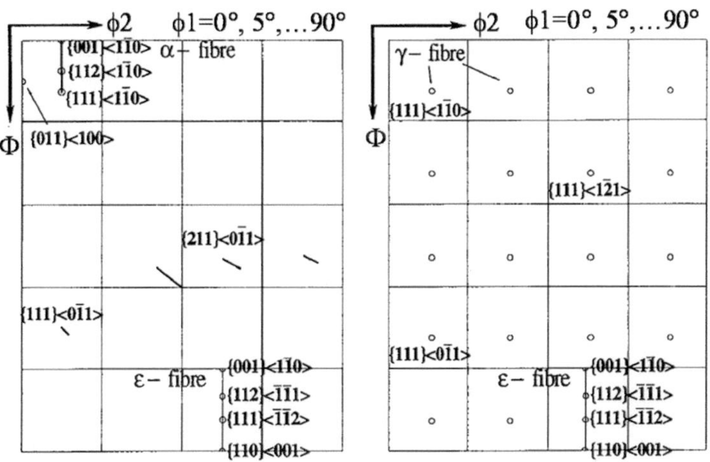

Figure 2.8: Schematic representation of important fiber texture components found in bcc metals [23]

In the left cross section, the α-fiber can be found as a line at φ1=0 and the ε-fiber appears as a line at φ1=90°. The right cross sections contain the γ-fiber that appears as single points in each cross section. From Figure 2.7 and Figure 2.8 it can be seen that the α-fiber lies in a section parallel to φ1=0 and φ2=45°. The γ-fiber can be found at 0°<φ1<90°, Φ=54.7° and φ2=45°. Therefore a section parallel to φ2=45° can be used to visualize the α–fiber and the γ-fiber. A scheme of a section parallel to $\phi2=45°$ with the most prominent orientations for bcc metals is shown in Figure 2.9.

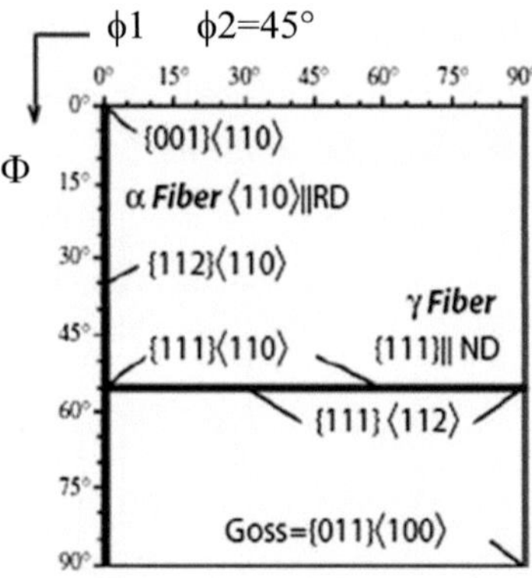

Figure 2.9: Schematic representation of a $\phi2=45°$ section with the most important orientations in bcc materials [9]

Raabe et al. [23] investigated the texture evolution of low carbon steels during cold rolling. They observed the development of a strong α-fiber and the formation of a γ-fiber. Dependent on the rolling degree, different texture components are dominant (see Figure 2.10). For $\varepsilon < 70\,\%$ an incomplete α-fiber between the {100} <110> and {112} <110> orientations exists and within the γ-fiber the {111} <112> orientations show a weak preference. The formation of the ε-fiber and the β-fiber could not be observed in these materials. For $\varepsilon > 70\,\%$ the maximum within in the α-fiber is found at {112} <110> and on the γ-fiber, the maximum moves to {111} <110> orientations. They found that in most low carbon steels the microstructure is homogeneous throughout the sample thickness.

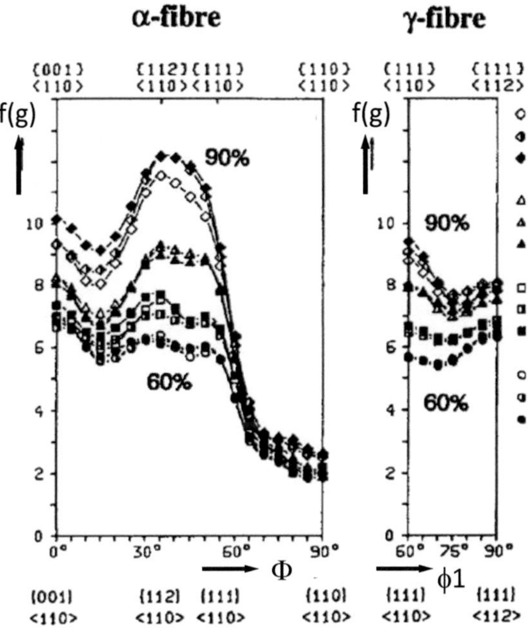

Figure 2.10: ODF plot for low carbon steel after different rolling reduction [23]

Von Schlippenbach et al. [27] investigated the evolution of the microstructure of four deep drawing steels. Armco Iron (AI) contains the same alloying elements as the DC04 steel investigated in this work at different concentrations. Figure 2.11 shows an ODF of a hot rolled vacuum degassed steel and the ODF of a AI steel after 80 % rolling reduction. The ODF of the hot band looks rather diffuse as the Euler angles are randomly distributed in the Euler space. However, after the cold rolling process a strong texture developed which is characteristic for all deep drawing steels that were investigated in this study. Weak differences can be found for the prominent orientations that developed in these four materials.

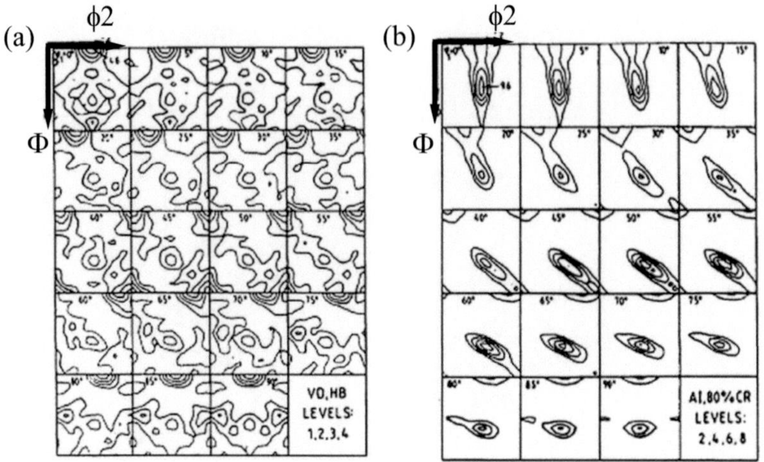

Figure 2.11: ODF of a (a) hot rolled vacuum degassed deep drawing steel and the ODF of a (b) cold rolled deep drawing steel with similar composition than the DC04 steel. [27]

The dominant orientations within the α-fiber and the γ-fiber can be found by plotting the orientation density *f(g)* along these two fibers (Figure 2.12). After 80 % rolling reduction the α-fiber ($\phi 1=0°$) developed in AI steel between {001} <110> and {111} <110>. Within the γ-fiber the {111} <110> orientations show the maximum intensity.

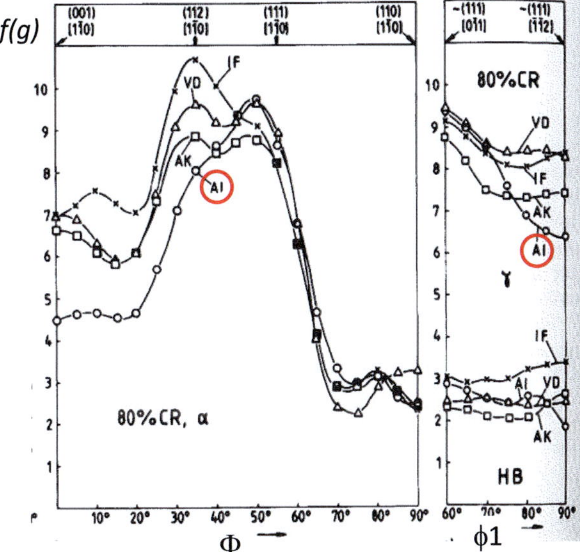

Figure 2.12: ODF plot of different deep drawing steels after cold rolling (80 %) [27]. The ODP plot on the left side shows the α-fibers and the ODF plots on the right side the γ-fiber.

2.2.3 Evolution of the microstructure during heat treatment

Metals that are exposed to deformation processes such as cold rolling contain high stored deformation energies due to dislocations and defects that were introduced into the crystal lattice. If the deformed material is subjected to sufficiently high temperatures, changes in the microstructure are observed while the dislocation density decreases. These microstructural changes can be divided into two major processes: recovery and recrystallization. Several theories exist that describe the mechanisms of dislocation motion during recovery, the nucleation of grains and the grain growth that take place during recrystallization. Therefore, detailed information about these mechanism can for example be found in the books of Gottstein [30, 31] and Humphreys and Hatherly [28]. In general, the processes that take place during recovery and recrystallization are based on the diffusion of atoms as well as glide and climb of dislocations. In this chapter, the mechanisms that control the recovery process, the nucleation of new grains and their growth will be addressed.

Recovery

In general, recovery takes place at lower temperatures or before the recrystallization process starts and is accompanied by the reduction of the dislocation density. After Humphreys and Hatherly [28], different processes occur during recovery, the annihilation of dislocations and the formation of small angle grain boundaries. Dislocations interact with each other due to their surrounding elastic stress fields. The interaction forces between two dislocations with Burgers vectors b_1 and b_2 is described by the Peach-Koehler equation

$$F = \frac{Gb_1b_2}{2\pi r_v(1-v)}\cos\Theta\cos2\Theta \qquad (2.6)$$

where r_v is the distance and Θ the angle between two dislocations. For $\Theta = 0°$ the dislocations are located on the same glide plane.

For two parallel dislocations that are located on the same glide plane, the force becomes positive and these dislocations repel each other. In contrast, for two antiparallel dislocations the force becomes negative and these dislocations attract each other and annihilate. In the case, that two parallel dislocations are located on parallel glide planes at an angle that is larger than 45°, the dislocations start to attract each other. In this case, the equilibrium position of these two dislocations is an angle of 90° which means that these dislocations are located above each other on parallel glide planes. This process leads to the formation of small angle grain boundaries during the recovery processes. Higher temperatures lead to increased mobilities of dislocations and promote diffusion which is necessary for dislocation climb, facilitating dislocation rearrangement and annihilation.

Recrystallization processes and their driving forces

Recrystallization can be divided into two mechanisms, the nucleation of new grains and a subsequent grain growth. In order to enable nucleation processes, different requirements need to be fulfilled which are described by Gottstein [31]. These requirements will be summarized here very briefly. From thermodynamics it is known, that the nucleus needs to have a critical size to ensure its stability. A further requirement for nucleation is an instability that leads to the motion of a grain boundary into a certain direction. Reasons for this can for example be an inhomogeneous dislocation density. A third requirement is a kinetic instability. A boundary around a nucleus needs to remain mobile (e.g. large angle boundaries). It can be very difficult to form a mobile boundary depending on the local microstructure. In order to nucleate grains, all of the three requirements need to be fulfilled which makes the nucleation process very site selective. Nucleation is expected to occur mostly at large angle grain boundaries and inhomogeneities of plastic deformation. Typically nucleation occurs simultaneously with recovery and is therefore very sensitive to the kinetics and the details of the local configuration of the defects.

During recrystallization, a new microstructure forms by the motion of large angle grain boundaries into the deformed material that leave behind an area with a reduced dislocation density. The process of nucleation and grain growth is described as primary recrystallization. During grain growth, two different cases can occur. First, the new grains that have developed during the nucleation process grow simultaneously into the deformed material until they impinge on each other. This process is coupled to an increase of the average grain size in which the standard deviation of the grain size distribution is kept constant and is therefore called continuous grain growth. In a second case single grains start to grow faster than the rest which leads to an increase of the average grain size but the shape of the grain size distribution is changing. Therefore, this process is called discontinuous grain growth or secondary recrystallization. Finally, the grain growth processes can finish when the size of a grain reaches the smallest dimension of the sample. In the case of a thin sheet metal, this dimension can for example be the sheet thickness. In this case, the grain growth is controlled by the surface energy. This process is called tertiary recrystallization.

In general, different driving forces affect the grain growth process which are summarized by Gottstein [25]. In order to enable the motion of the grain boundaries by a small distance Δx, the free energy of the whole system needs to be reduced. A driving force that controls the primary recrystallization process is the reduction of the stored deformation energy that was generated by dislocations in the crystal lattice. A grain that starts to grow into a region with high dislocation density by leaving a region with a reduced dislocation density can minimize the stored deformation energy within this region. If the material is strongly deformed, the dislocation density can lie in the range of 10^{15}-$10^{16}\mathrm{m}^{-2}$ and can reach values of roughly 10^{10}-$10^{12}\mathrm{m}^{-2}$ after recrystallization. The deformation energy per volume [J/m^3] is proportional to the density of dislocations ρ, the shear modulus G and the Burgers vector b:

$$E_{dislocation} = \frac{1}{2}\rho G b^2 \tag{2.7}$$

A second driving force during grain growth is the grain boundary energy that changes if the grain boundaries that start to grow into the deformed material. Assuming a grain with the shape of a sphere, the grain boundary energy per volume can be estimated by the pressure that a sphere feels due to the surface energy. The grain boundary energy can be calculated by dividing the derivative of surface (interface) energy with respect to radius by the surface area of the spherical grain.

$$E_{grain\ boundary} = \frac{d(4\pi R^2 \gamma)/dR}{4\pi R^2} = \frac{8\pi R \gamma}{4\pi R^2} = \frac{2\gamma}{R} \qquad (2.8)$$

This result is not very different from the ratio of surface energy per volume of the sphere $(4\pi R^2 \gamma/(4/3\pi R^3) = 3\gamma/R$.

A third driving force that may influence the recrystallization process is the elastically stored energy. This energy depends on the elastic strain energy in a region of two neighboring grains with elastic moduli E_1 and E_2. The overall energy can be reduced when the grain with the lower stresses grows in expense of the grain with the higher stresses.

$$E_{elastically\ stored} = \frac{\sigma^2}{2} \cdot \left(\frac{1}{E_1} - \frac{1}{E_2} \right) \qquad (2.9)$$

In order to compare the three different driving forces, typical energies were calculated dependent on the dislocation density and the curvature of the grains. This calculation was performed within the diploma thesis of Moritz Wenk [32] where the microstructure evolution of cold rolled DC04 steel was investigated during a heat treatment. Table 2.2 summarizes the values that were chosen for the calculation of the three energies that control the recrystallization process. The results are plotted in Figure 2.13. The estimations show that for grains with radii (grain sizes) smaller than 0.1 µm and low dislocation densities, the grain boundary energy becomes the dominant term in the recrystallization process. For high dislocation densities and grain radii that are larger than 0.1 µm, the plastically stored energy controls the annealing process. The elastically stored energy becomes dominant in the case that the grains are relatively large and the dislocation density is low.

Deformation energy	$E_{dislocation} = \dfrac{1}{2}\rho G b^2$	$\rho=10^{12}-10^{16}\text{m}^2$ $G=82.4\text{GPa}$ $b=2.86\text{ Å}$
Grain boundary energy	$E_{grain\ boundary} = \dfrac{2\gamma}{R}$	$\gamma=1\text{J/m}^2$ $R=10^{-8}-10^{-3}\text{ m}$
Elastically stored energy	$E_{elstically\ stored} = \dfrac{\sigma^2}{2}\cdot\left(\dfrac{1}{E}\right)$	$\sigma=1000\text{ MPa}$ $E=206\text{ GPa}$

Table 2.2: Values that were assumed in order to compare the different energies that control the recrystallization processes. The elastic stresses were determined within the diploma thesis of Moritz Wenk. The other values were received from the literature [31]

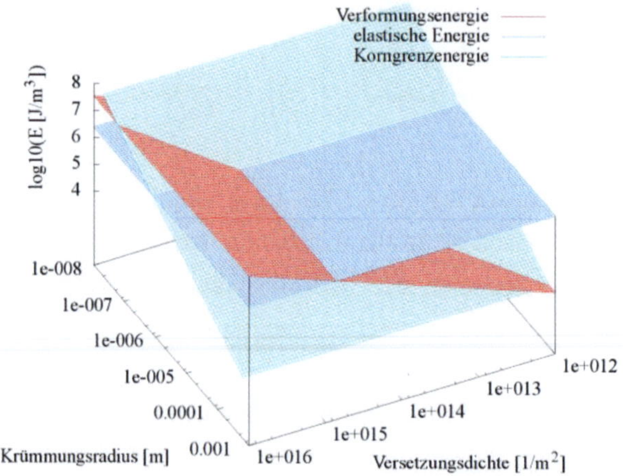

Figure 2.13: Estimated deformation energies, grain boundary energies and elastically stored energies that typically control the recrystallization processes. (calculated by Moritz Wenk [32])

2.3 Strengthening mechanisms and size dependent deformation behavior of metals

2.3.1 Strengthening mechanisms in metals

The deformation behavior of metals is mediated by the activation of dislocation sources, dislocation motion and dislocation interaction. If the dislocation density and the Peierls energy are rather low, the dislocations can be activated easily which leads to relatively low strengths of the material. Different mechanisms exist that can hinder the dislocation motion and increase the strength which are discussed in detail in the book of Courtney [33]. Typical examples are substitutional [34, 35] or interstitially incorporated atoms into the crystal lattice. Examples for interstitially dissolved elements are C and N in Fe that cause a tetragonal distortion of the crystal lattice. This tetragonal distortion leads to an interaction between the atom and the stress field of the dislocation. The increase of the strength $\Delta\tau$ is proportional to the maximum interaction force F_{max} that is needed to pass an obstacle and inversely proportional to the distance L_p between the obstacles. In the case of interstitially dissolved atoms, the average particle distance can be estimated by the inverse square root of the carbon concentrations $c^{-1/2}$. The increase in strength $\Delta\tau$ caused by the alloying elements is then proportional to the shear modulus G and the square root of the carbon concentration $\sqrt{c}$.

$$\Delta\tau \sim G \cdot \sqrt{c} \tag{2.10}$$

This $\sqrt{c}$ dependency was investigated by Wert et al [36]. In their studies. They showed that a carbon concentration of only 0.0002 wt% increases the flow stress by roughly 130 MPa.

Other mechanisms that can hinder dislocation motion are particles that are dispersed in the material. If these particles are large and their lattice incoherent with the lattice of the surrounding material, the dislocations cannot bypass them. Instead, a dislocation has to bow around the particle and leaves a dislocation loop that surrounds this particle. This is the Orowan mechanism [37]. The stress that is required to bypass the obstacle is inversely proportional to the particle spacing.

$$\Delta\tau \cong Gb/L \tag{2.11}$$

A subsequent dislocation has to bypass the particle and additionally the surrounding dislocation. This reduces the effective particle distance and therefore increases the stress that is required to bypass the particle and the residual dislocation loops.

With increasing amount of deformation, the dislocation density ρ increases within the metal. These dislocations can also act as obstacles and can impede the dislocations motion. Each dislocation is surrounded by a stress field that is caused by a distortion of the lattice. The stress fields of the dislocations interact with each other and can lead to repulsion or attraction. Moving dislocations interact and form intersections which can lead to immobile jogs. This leads to obstacles for further motion. In this case, the obstacle distance can be estimated by the inverse square root of the dislocation density $\rho^{-1/2}$. The increase in strength is then proportional to the square root of the dislocation density ρ which is described by the Taylor equation [37]

$$\Delta\tau = \alpha Gb \cdot \sqrt{\rho} \tag{2.12}$$

where α is a dimensionless constant in the regime 0.1 to 1. During deformation, dislocations often form networks [38, 39] with characteristic spacings in the micrometer range. These dislocation networks also hinder dislocation motion and increase the strength.

For many metals, a strong increase in flow stress and fracture toughness can be observed with decreasing grain size. During the deformation of these metals, moving dislocations start to pile up at grain boundaries which act as obstacles for the moving dislocations. The stress fields of the dislocations in the pile up at the grain boundaries overlap and increase the force on the boundary. For larger grains more dislocations pile up and higher effective stresses can act on the grain boundary for a given stress in a grain. Therefore higher stresses are required in smaller grains to initiate flow in the neighboring grain. The increase in stress is inversely proportional to the square root of the grain size D [40].

$$\sigma_y = \sigma_0 + \frac{k_y}{\sqrt{D}} \tag{2.13}$$

This is the Hall-Petch relation which is valid as long as the grains are larger than a critical size which depends on the material. This critical grain size is usually in the range of several tens of nanometers. In nanocrystalline materials that have grain sizes below this critical value softening effects may occur [41].

These examples show that the increase in strength of metals scales with the spacing of obstacles that impede dislocation motion. Within the microstructure, characteristic length scales can be particle spacings, dislocation spacings or grain sizes [37]. Besides microstructural constraints, the strength of the material can also be enhanced by a reduction of the sample dimensions.

2.3.2   Mechanical size effects due to dimensional constraints

With the rise of the microelectronics industry, interest in the mechanical behavior of thin films and small conductor lines became of great interest. Examples are thin films of Cu [42] and Al [43] with thicknesses of several μm that are deposited onto a substrate that is much thicker than the film itself. Nix [44] developed a model for single crystalline films which quantifies a biaxial flow stress σ in the thin film that is proportional to the inverse film thickness. Until today, strong efforts are devoted towards developing an understanding of how the strength of these material systems is affected by film thickness or sample size, the grain size and alloying elements. In the following, relevant results from the literature will be summarized for fcc and bcc metals. This information serves as background for the part of this thesis where the mechanics of small steel samples are investigated (Chaper 5).

A technique that is frequently used to investigate the deformation behavior of samples with sizes in the micrometer range, is the micro compression technique [7]. It can be used to investigate pillars with diameters as small as ~100 nm. Details on this method will be given in section 3.3.1. This method is very flexible and important for this thesis since it allows site specific experiments in selected regions of the sample. A large amount of data and understanding of small scale behavior has been acquired using this method. Results that were obtained mainly by this but also by other methods suggest the following scaling behavior of small-scaled metals. In general, the increase in strength with decreasing pillar diameter can be described by a phenomenological relationship between the yield strength σ_y, the bulk yield strength σ_0 and the pillars diameter d,

$$\sigma_y(d) = \sigma_0 + c \cdot d^{-\beta} \tag{2.14}$$

The exponent β describes the scaling of the yield strength with the pillar diameter. This equation represents a general form to describe size effects. For $\beta = 0.5$ it represents the Hall-Petch relation and for $\beta = 1$ the Nix model for thin films.

FCC metals

In fcc metals, dislocations move on $\{111\}$ planes and glide occurs along the $<110>$ direction which results in 12 different glide systems. During deformation, the glide systems that experience the highest resolved shear stress are activated. This is described by the Schmid's law. In order to activate glide processes in metals, a critical resolved shear stress (CRSS) has to be overcome. The resolved shear stress that is acting on a glide system can be calculated by projecting the loading stress σ_0 that is applied on the pillar surface onto the glide systems. As long as the angle between the loading direction and the glide direction is between 0 and 90° the glide system experiences a shear component. This shear component can be evaluated by multiplying the loading stress by the Schmid factor. The Schmid factor is determined by

$$CRSS = \sigma_0 \cdot \underbrace{\cos\lambda \cdot \cos\emptyset}_{\text{Schmid factor}} \tag{2.15}$$

$\emptyset$ is the angle between the normal vector of the glide plane and the loading direction and λ is angle between loading direction and the glide direction.

In fcc metals the size dependent deformation behavior for different materials such as Cu [45], Au [46-49], Ni [7, 50-52] and Al [53] has been measured for pillar diameters ranging from 0.2 µm to 22 µm. The orientations of the pillars range from single slip orientations to multiple slip orientations. Commonly, the stress-strain response is calculated from the recorded force-displacement data. Dependent on the pillar sizes, the stress-strain behavior can differ significantly. Pillars with diameters larger than ~1 μm show a smooth hardening behavior as it is known from bulk materials. The stress strain data for pillar diameter smaller than ~1 μm exhibit strain bursts which are intervals without any strain hardening [7] that occur after a critical strain and happen randomly. In order to compare different materials and different pillar sizes, the flow stresses are plotted for the corresponding pillar diameter. Dependent on the authors, flow stresses have been taken at different strain of 2%, 2.5%, 5% or 10% and the diameters were either measured at the top or at the mid-height of the pillar. Despite these differences, the data can be compared to obtain a scaling behavior. The flow stresses are normalized by the shear modulus and the Burgers vector and plotted in a double logarithmic plot versus diameter or sample size [54, 55] according to equation (2.14). When applying this procedure to the flow stress data for a collection of fcc metals, the data follow a trend line with a slope of $\beta \approx 0.6$. [49, 56, 57]

BCC Metals

During the last years, studies have been carried out in order to investigate the size dependent deformation behavior of bcc metals including Ta [58], W [59], Mo [59-61] and Nb [59, 62]. The microcompression experiments that were performed on these metals show a pronounced size dependent behavior with exponents ranging from -0.21 to -0.48 [59]. Schneider et al. [59] suggested that the rather weak size scaling is a consequence of the limited mobility of screw dislocations in bcc metals and assumed that the size effect exponent might depend on the athermal temperature of the material. In bcc metals, the mobility of screw dislocations is sensitive to temperature and for sufficiently low temperatures dislocation motion is only mediated by the nucleation and motion of kink pairs. With increasing temperature the mobility of screw dislocations increases and at the athermal temperature the mobility of edge and screw dislocations eventually are equal [63].

For α-Fe the athermal temperature of ~300 K [64-67] is close to room temperature. Kaufmann [56] investigated the size dependent deformation behavior of α-Fe by microcompression experiments at room temperature (Figure 2.14a). Figure 2.14 b shows the stresses at 5% plastic strain plotted for the corresponding pillar diameter. When describing the scaling by an exponential relationship an exponent of -0.81 is found. According to this, the high mobility of screw dislocations should lead to a size scaling similar to fcc metals. This coincides with the strong size dependence observed here. The fact that the size dependence (Figure 2.14b) has an exponent of -0.81 relative to values of -0.6 that are commonly found for fcc metals may be due to fundamental differences between fcc and bcc metals or the peculiar magnetic interaction that is known to complicate plastic flow of α-Fe [68, 69]. Compared to α-Fe, the athermal temperature for W and Mo lie at 800 K [70] and 480 K [71] respectively. Therefore, it is argued that the lower exponents compared to α-Fe are related to the low mobility of screw dislocations.

Differences in the stress-strain response are also apparent from the data in Figure 2.14a. Large α-Fe pillars with diameters around 4 μm showed a deformation behavior similar to what is found for bulk materials. After elastic loading, plastic flow sets in and during further deformation hardening was observed. When comparing curves from different experiments, little scatter in the stress-strain behavior is found. With decreasing pillar size, the strength increased and the stress-strain curves became increasingly disruptive due to strain bursts. The small pillars showed more variations in the stress-strain response and a larger scatter in their strength.

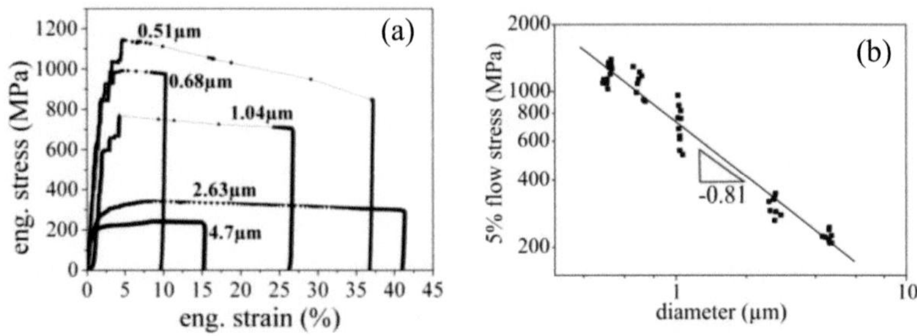

Figure 2.14: Stress-strain data of single crystalline [211] α-Fe pillars with diameters ranging from 0.51 μm to 4.7 μm (a) and the flow stress for the corresponding pillars diameters at 5% plastic strain (b). An exponential relationship with an exponent of -0.81 is used for describing the data. [56]

The effect of pre-straining on the size dependent deformation behavior was investigated by Bei et al. [72]. They presented a new technique for preparing almost dislocation free Mo fibers that were grown in a NiAl-Matrix by a eutectic solidification process. The pillars were produced by an etching process and therefore no FIB machining was necessary. In TEM studies they showed that these pillars are almost dislocation free. They argued that the low density of defects leads to a nucleation controlled behavior and high stresses. In the experiments, stresses of almost 10 GPa were achieved during micro compression, with no size dependence. Some of the samples were plastically prestrained up to 11 %. For pre-strains up to 4 % the stress strain data showed a stochastic behavior with strain bursts and for 11 % pre-strain the stress strain behavior changed to a bulk like behavior with continuous work-hardening of the sample. Overall, after pre-straining the strength of the pillars decreased and a size dependent strength could be observed. From this observation it was concluded that the non-existing size dependent deformation in pillars without pre-strain is an effect of the lack of dislocations.

Mechanisms that may explain mechanical size effects in micro pillars

Different mechanisms were suggested that may cause the observed size dependent behavior. In bulk material, strain hardening occurs by the interaction and multiplication of dislocations. Dislocation networks with a characteristic length scale of a few micrometers form which hinder dislocation motion. This is different when the sample size becomes smaller than the characteristic length scale of the dislocation networks. In this case, the strength of the material cannot longer be controlled by the interaction of the dislocations with the network and the interaction between individual dislocations becomes increasingly important. With further decreasing sample size, the number of available dislocations in the sample decreases as well as the number of available dislocation sources. This is a consequence of the decreasing volume. Besides the volume, the surface area decreases but the surface area to volume ratio increases. This leads to a higher probability for the dislocations to annihilate at the nearby surface and may lead to a state where no mobile dislocations are present. This effect is called "dislocation starvation" [47]. Other models consider the reduced diameter of the sample which limits the size of dislocations sources, such as single-armed Frank-Read sources. Therefore, the critical stress to activate these sources scales inversely with the sample size [73]. Also, the likelihood of having a suitable dislocation source decreases [74, 75].

3. Experimental

3.1 Microstructural investigation methods

3.1.1 Dual Beam SEM and FIB

Focused ion beam microscopy (FIB) and secondary electron microscopy are conventional techniques that are widely used for the investigation of small features down to the nanometer scale. In this work, all microstructural investigations were performed with a dual beam microscope (Nova Nanolab 200, FEI, Hillsboro, Oregon 97124, USA) which is equipped with a SEM column and a Ga^+ ion column that is mounted at an angle of 52° with respect to the electron column. The electron column is equipped with a thermal field emission gun. The maximum resolution that can be achieved is ~1 nm. Typical acceleration voltages are on the order of several kV. In this work, mostly a voltage of 10 kV and a beam current of 0.54 nA were used for imaging. For EBSD 20 kV and 6.6 nA were used. In general, higher acceleration voltages lead to a deeper penetration of the electrons which results in secondary electrons that are coming from deeper regions of the sample. The secondary electrons that leave the sample are detected and used for imaging. Secondary electrons have lower energies (< 50 eV) than the incoming electrons and are very sensitive with respect to the surface topography. Besides secondary electrons, backscattered electrons are generated within the sample that can be used for sample imaging. The number of backscattered electrons depends on the number of electrons of an atom and consequently on its atomic number. Therefore, different phases appear as regions of higher and lower intensities in the SEM images. Besides imaging, these backscattered electrons can be used to analyze the crystal structure which will be explained in detail in section 3.1.2.

In the FIB, Ga^+ ions are emitted and a voltage of 30 kV accelerates the ions towards the sample. The Ga^+ ions penetrate the sample and generate electrons that can be used for imaging. The imaging resolution is approximately 7 nm but with increasing beam current the resolution strongly decreases. Dependent on the application, Ga^+ currents between 10 pA ("non destructive" imaging) and 20 nA (strong milling) are used. Ga^+ ions that impact the sample surface can be used for sample milling. The sputter rate depends on the sample material, the acceleration voltage and the beam current. In this work, the Ga^+ ions were used in selected regions for the preparation of micropillars (see section 3.3.1).

3.1.2 *Electron Backscatter Diffraction (EBSD)*

The dual beam microscope that is described in section 3.1.1 is equipped with an EBSD detector that provides the possibility to determine crystallographic orientations. The EBSD detector consists of a phosphor screen with a Charge Coupled Device (CCD) camera on its backside. Inside the SEM chamber the sample is tilted to an angle of 70° (see Figure 3.1). The electron beam is focused to a point on the sample surface and the electrons are diffracted at an angle of θ on crystallographic planes that fulfill Braggs Law [8]. The diffracted electrons that emit the sample hit the phosphor screen and form diffraction patterns on the phosphor detector. These patterns consist of Kikuchi bands that can be recorded by the CCD camera. The symmetry of the crystal structure is reflected in the Kikuchi patterns and the three dimensional crystal orientations can be calculated by analyzing the patterns. The EBSD detector used in this work was manufactured by HKL, now Oxford Instruments (Tubney Woods, Abington, Oxfordshire, OX13 5QX, UK) and the recorded Kikuchi pattern were analyzed using a computer program form the same company named

"Channel 5". The crystal orientation is usually expressed in terms of Euler angles ϕ_1, Φ and ϕ_2, which describe the crystal coordinate system with respect to the sample coordinate system. As described in section 2.1, the Euler angles are used in the Bunge convention.

The pattern quality and therefore the detection efficiency strongly depends on the quality of the sample surface and in order to obtain sample surfaces that are smooth enough for this method, the DC04 specimens were prepared carefully. They were first mechanically ground with SiC paper of decreasing grain size and polished with 6 µm, 3 µm and 1 µm diamond paste. In order to remove remaining deformations that were introduced by the mechanical polishing, in a last step the samples were electrochemically polished using a Struers D2 electrolyte.

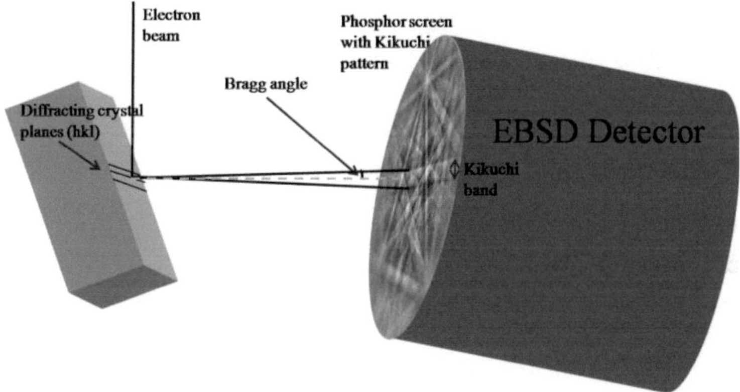

Figure 3.1: Schematic EBSD set-up and principle of Kikuchi pattern formation

3.2   Material: DC04 Steel

The DC04 steel is a standard deep drawing steel that is for example used in the automotive industry. The material used in this work was produced by Voestalpin AG (Linz, Austria). According to the material specification, DC04 steel contains 0.08 wt% C and 0.40 wt% Mn as well as other impurities like S and P at lower levels [6, 76]. All DC04 samples originated from the same batch and were extracted after different process steps which were hot rolling, a subsequent cold rolling and a final heat treatment step. These processes led to changes in the microstructure and the mechanical properties that were investigated on different length scales and for different sample geometries.

3.3   Sample fabrication

3.3.1   Pillar preparation methods using FIB

Besides imaging, the FIB has been used to mill small samples with sizes in the micrometer range into the material. Pillars with sizes of several micrometers in diameter and height can be made within several hours. Today, this method has been widely established for preparing samples for small scale testing. Problems arise from the Ga^+ ions that enter the sample and can lead to heavily damaged surface layers that may have an impact on the deformation behavior of the material. The implantation depth of the Ga^+ ions into the sample surface can be estimated using the SRIM code [77] (Stopping Range of Ions in Matter). Calculations for polycrystalline Cu showed that the penetrations depths can reach 50 nm for an acceleration voltage of 30 keV. The intercalation of the Ga^+ ions into the sample can cause problems that have been discussed by several authors [78-81]. The implantation of Ga^+ ions can cause point defects and dislocations that form dislocation networks and affect the strength of the material. These defects can lead to a softening of the metal [80]. In another case, an amorphized surface layer was observed [79]. There the opposite effect can happen. Moving dislocations pile up at the surface layer which increases the strength of the tested metal. Although this method has limitations, it offers the possibility to machine pillars into selected regions of the sample which is of interest particularly for site specific tests in polycrystalline or composite materials.

Top down milling

A method to produce several pillars within reasonable time is the so called top down milling method. For this procedure the sample stage is tilted to an angle of 52° so that the sample surface is aligned perpendicular to the incident Ga^+ beam. This process consists of two steps, first a coarse cut with a high beam current to remove most of the material around the pillar. A circular pattern containing the surrounding trench of the pillar is milled into the sample in multiple passes from the outer edge to the inner edge of the trench. Afterwards a fine cut with a small beam current is applied as the final step. In this step, the final pillar diameter is reached in only one pass so a dwell time of several milliseconds is necessary. The acceleration voltage of the ion beam was 30 kV and the ion current depends on the process steps, pillars sizes and the material. Coarse cuts are usually performed with ion currents of 7 nA or 20 nA and the currents for fine cut are in the range of 0.5 nA and 3 nA. The top down milling method is useful for pillars with diameters of up to 8 µm but for larger pillars this method becomes very time consuming and even just the coarse cuts require more than one day. The disadvantage of this method is that the pillars do not have a perfect shape and have a taper ~5°. A method that prepares cylindrical shapes is the so called Lathe technique.

Lathe technique

The lathe technique can be used for different sample sizes. In general, tapered pillars are made first and then lathe cutting is used to remove the taper. In this work, lathe cutting was applied to clean the surface of very large pillars (Figure 3.3a and Figure 3.5). Lathe cutting consists of multiple steps that can be automated by using a FEI AutoScript software installed on the microscope. On the top of the pillar a small circle is milled which is used as a reference point to position the sample before each milling step using image recognition. Inside the FIB chamber the sample is tilted to -10° so the ion beam has an incident angle of 28° relative to the sample surface. A pattern with the shape of an L is milled at the edge of the pillar and afterwards the sample is rotated around the pillar axis by 6°. This procedure is repeated until the sample is rotated around 360° and in total and has reached its original position. The overlapping L shaped cuts then define the trench around the pillar. This procedure takes around 3-4 hours. Therefore only a few pillars were prepared by this method. The image of such a pillar (Figure 3.5) shows that the pillar has almost no taper on its sidewalls.

3.3.2 DC04 Micropillars

Pillars of different sizes of up to 6 μm were milled into individual grains using FIB. In this way pillars with different diameters were prepared in grains with <111>, <123> and <001> out of plane orientations. The pillars were divided into the according three classes with <111>, <123> and <001> orientations along their axes where misorientations from the nominal direction of up to 10° were tolerated. Figure 3.2a shows a typical single crystalline pillar that was machined into the heat treated material.

Besides the experiments on single crystalline pillars, polycrystals were tested in order to investigate the effect of grain boundaries on the mechanical response of the heat treated DC04. In general, five parameters are required to describe a boundary [82] (the orientation relation between the neighboring grains and the relative orientation of the plane of the boundary) , therefore many possible configurations exist and a systematic study would be very time consuming. This was not attempted in this thesis, instead boundaries were randomly selected and polycrystals such as the one shown in Figure 3.2b were produced and mechanically tested.

The effect of predeformation on the mechanical response of DC04 steel was also investigated by microcompression experiments using the deformed cold rolled DC04 steel. Pillars with diameters of 500 nm, 1.5 µm and 4 µm were micro machined by FIB into the cold rolled material. The orientation of the pillars could not be determined because the pillars were polycrystalline and had large misorientations as can be seen in Figure 3.2c.

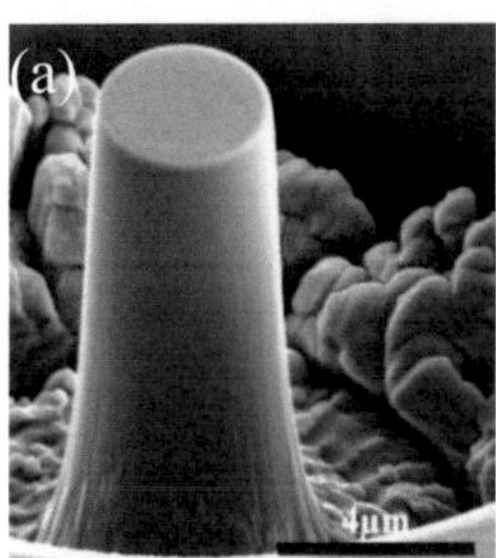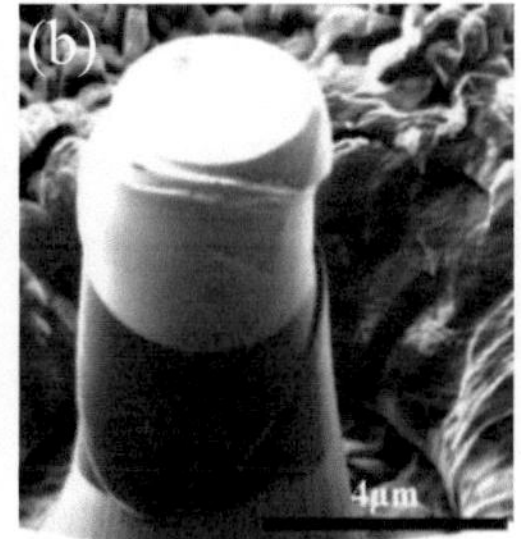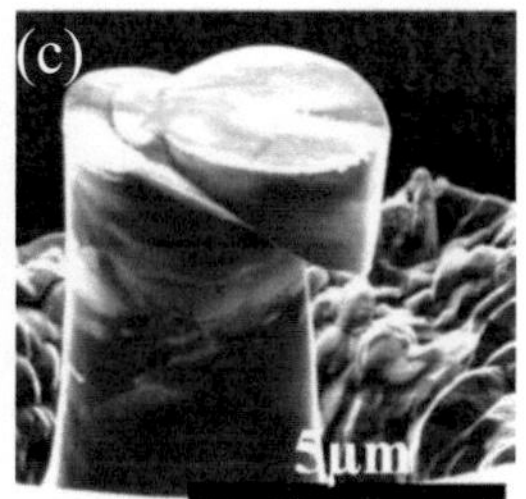

Figure 3.2: SEM images of pillars micro machined into the heat treated and cold rolled material.

For pillars with diameters larger than 10 µm, the time for the FIB machining, which scales with the volume of the pillar, is often too long and electrical discharge machining (EDM) was used instead. The coarse manufacturing of pillars with a diameter of 22 µm was performed by wire EDM in a commercial system. A thin brass wire and the sample are located inside a dielectric (water) and an electrical field is applied between both electrodes. A spark discharge between the wire and the sample is used to melt both surfaces and to remove material from the sample. Figure 3.3a shows an image of a pillar that was micro machined in the heat treated DC04 steel with a diameter of 22 µm and a height of 60 µm (cooperation with Christoph Ruhs, WBK, KIT). Unfortunately, EDM leads to relatively rough surfaces which had to be removed by FIB. Figure 3.3a shows the image of a pillar that was machined by EDM and Figure 3.3b a cross section prepared into the heat treated DC04 that was treated by EDM beforehand. The remaining rough surface had to be removed because it can be expected that it affects the deformation behavior of the pillars for example due to the different mechanical properties of this layer. The removal occurred by the lathe method as described in section 3.3.1.

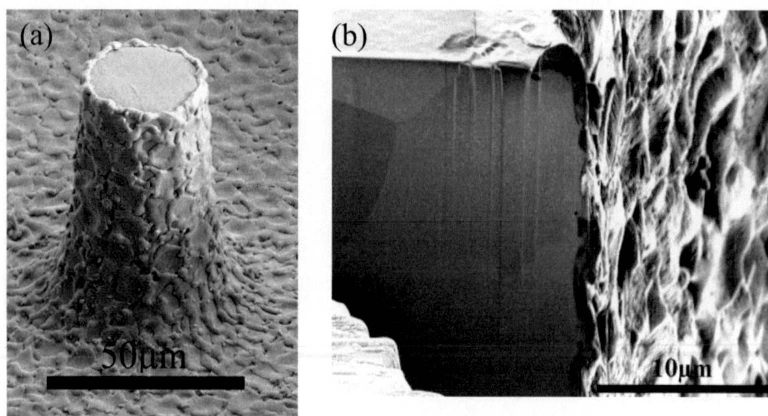

Figure 3.3: (a) Pillars prepared by EDM in the heat treated DC04 steel, (b) cross section of a solidified layer

3.3.3 DC04 samples for heating experiments

Two heating experiments were performed with the cold rolled DC04 steel in order to investigate the evolution of the microstructure during different heating conditions. These experiments were jointly performed as part of a diploma thesis of Moritz Wenk [32]. Small samples of roughly 0.25 cm^2 were prepared from the sheet metal using a metal saw and electropolished afterwards using a Struers A2 electrolyte.

3.3.4 DC04 samples for tensile experiments

Tensile specimens with a gauge length of 15 mm, a width of 5 mm and a thickness of 1.2 mm were produced by EDM. These samples were oriented parallel to the rolling direction as well as 45° and 90° to the rolling. Further tensile specimens with a gauge length of 7 mm and a width of 1.2 mm were also produced by EDM. These samples were oriented parallel to the rolling direction. After the rolling process, the material had a thickness of 1.2 mm which was too thick for the grip holder of the tensile machine (Zwick/72.5). To solve this problem, the samples were ground down to a thickness of 0.9 mm from the backside. After the grinding process both surfaces were electrochemically polished with a Struers A2 electrolyte to obtain a smooth surface as required for EBSD measurements.

3.4   Mechanical characterization methods

3.4.1   Tensile experiments

The samples described in section 3.3.4 were loaded parallel, 45° and 90° relative to the rolling direction in order to investigate the anisotropic deformation behavior of the material. The experiments were performed with a tensile machine (Instron 4504) at a strain rate of 1 mm/min to total plastic strains of up to 20%. Additional tensile experiments were performed on smaller samples that have gauge lengths of 7 mm, widths of 1.2 mm and thicknesses of 0.9 µm. These sample samples were aligned parallel to the rolling direction and the experiments were performed with a tensile machine (Zwick/72.5) at a strain rate of 4 µm/s. The first sample was strained to fracture and the second sample was unloaded after 1 %, 5 %, 11 %, 19 % and 28 % residual plastic strain. The second sample was transferred to the SEM after each unloading step to determine the crystal orientations by EBSD.

3.4.2   Compression of micro pillars using a Nanoindenter

Mechanical compression experiments are a useful technique to measure the mechanical behavior of samples with sizes down to ~100 nm. These experiments were all performed in a Nanoindenter XP (MTS System Corp., Eden Prairie, MN, USA). Figure 3.4 shows a schematic of a nanoindenter. The indenter is composed of a coil-magnet assembly that controls the force that is applied to the sample. The displacement of the indenter tip is measured by a capacitance gauge consisting of a three plate capacitor. The springs keep the indenter shaft in the lateral position. Dependent on the experiments, different tips such as Berkovich indenters, spherical indenters, cube corner tips or flat punches can be mounted to the shaft. The sample is mounted on a xy-stage. The indentation locations can be selected by moving the sample stage under an optical microscope. A more detailed description of performing indentation measurements can be found in [83].

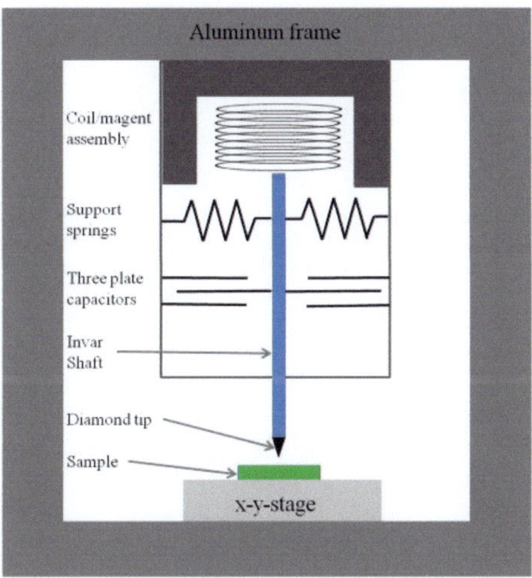

Figure 3.4: Schematic assembly of a Nanoindenter XP (MTS)

For the micro compression experiments, indenter tips with the shape of truncated cones were used. The sidewalls of these indenter tips have angles of 60° and depending on the pillar sizes, the flat punches have diameters of 10 μm or 100 μm. Figure 3.5 shows a schematic of a pillar and the flat punch tip. The experiments were performed under load control, therefore a constant loading rate was applied during the experiments. The pillars were compressed using a stress rate of ~20 MPa/s to a nominal displacement of 10 % of the original height.

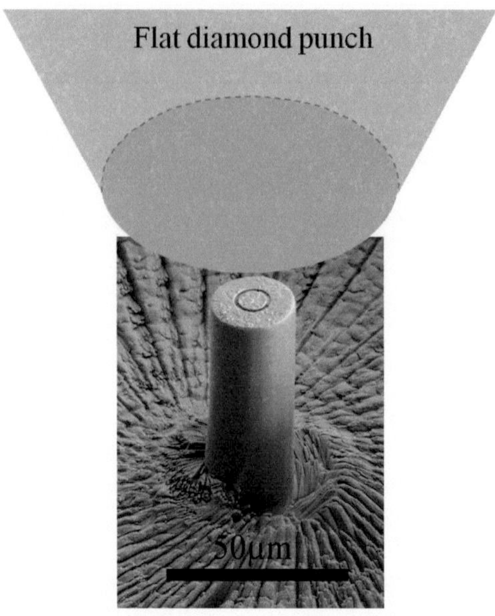

Figure 3.5: Pillar micro machined with a FIB and a schematic of a compression test. The pillar in the image is the one depicted in Figure 3.3a after cleaning by the Lathe technique (section 3.3.1).

Although microcompression experiments are a versatile technique for mechanical testing of samples in small dimensions, this method has limitations that have to be discussed [54]. Usually the stress distributions in the pillars are not homogeneous and fully uniaxial due to several reasons [84].

- The friction between the pillars and the punch causes stresses concentrations at the top of the pillars. This can lead to localized deformation events at the top of the pillar but can also help to stabilize the pillars and can avoid bending problems.
- The non-perfect geometry of the pillars (taper) leads to a non-uniform stress distribution along the pillars.

- Misalignment between the pillars and the punch are possible and have an influence of the deformation behavior of the pillar
- Additional stress concentrations will occur at the root of the pillar which can affect the load displacement response of the sample.
- As already discussed, FIB machining leads to an implantation of Ga^+ ions into the sample which can affect the mechanical responses of the samples (section 3.3.1).
- Buckling of the pillar can be an issue. To minimize such effects, the aspect ratios of the pillars were chosen to be between two and three (a compromise between avoiding buckling and excluding strong effects from the root/support of the pillar).

4. Texture and Microstructure of DC04 Steel after Cold Rolling and Heat Treatment

The process chain of DC04 steel is composed of hot rolling, cold rolling and a final heat treatment process. All of these processes have some impact on the microstructure and therefore affect the mechanical properties of this material. The aim of the Graduiertenschule 1483 is to reveal and simulate the interconnection between the processing conditions and the mechanics of DC04 steel. In order to achieve this aim, it is instructive to start with the different processing conditions and to observe how they affect the microstructure as a first step before further investigating how the different microstructures then relate to different mechanical properties of the resulting material. This chapter reports on the microstructures that exist after different processes with a main focus on the phenomena that underlie the heat treatment step. In section 4.1, the microstructures of DC04 steel after hot rolling, cold rolling and a heat treatment were investigated by EBSD. In section 4.2, results from slow annealing experiments on cold rolled DC04 steel are presented. In these experiments, the evolution of the microstructure was monitored by EBSD after different heating steps to track the development of the microstructure with focus on grain nucleation and growth. In section 4.3, the influence of a higher heating rate on the microstructure development is addressed. Microstructures from the experiments in section 4.2 and section 4.3 are also compared to the microstructure of DC04 taken directly after the heat treatment in the production process. The differences and possible mechanisms that are responsible for these differences are discussed in section 4.4.

4.1 Texture and microstructure of DC04 steel after hot rolling, cold rolling and a heat treatment process

DC04 steel was extracted from the production chain after the hot rolling, the cold rolling or the subsequent heat treatment process. These materials are the reference materials for this thesis and in this section their microstructure was investigated by SEM, FIB and EBSD. Figure 4.1a-c show the grain morphologies of these samples. Figure 4.1a and c were imaged by FIB which can be used to visualize different grains due to a channeling effect of Ga^+ ions in the crystal lattice. This channeling mechanism is based on the fact that ions that penetrated deeply, produce less detectable electrons. For example grains with the densely packed <111> planes appear brighter (more electrons) than <100> planes. The SEM image in Figure 4.1b shows the grain morphology of the cold rolled material. On very smooth surfaces, contrast of backscattered electrons can be used to visualize different grains due to the fact that standing electron waves form, depending on the crystal orientation which lead to different backscatter for different orientations. This so called electron channeling is different and more angle selective than the channeling of ions, which results from ion-solid interactions. Despite this difference, both effects can be effectively used to visualize grains.

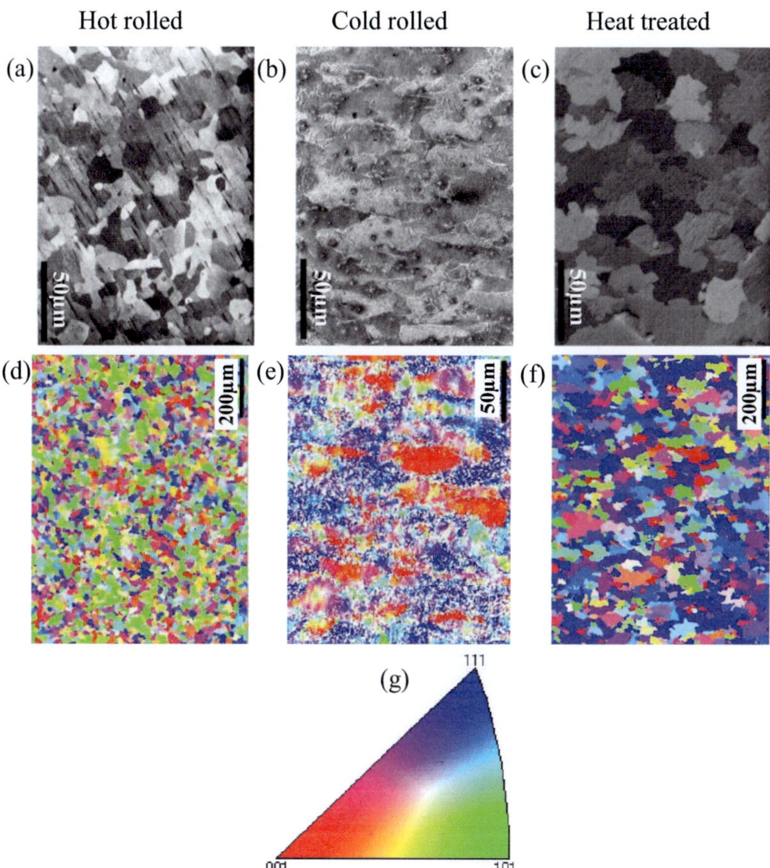

Figure 4.1: (a) and (c) FIB images of the rot rolled (a) and heat treated (c) DC04 steel. Figure (b) SEM image of the cold rolled DC04 steel. The EBSD results (d) – (f) show the crystal parallel to the sheet normal, determined form the samples that are depicted in (a) – (c). Since the different states have microstructures of different length scales, data were recorded at different magnifications. The colors in the EBSD maps correspond to the color coded orientations in the inverse pole figure map (g)

In order to obtain information on the crystal orientations, EBSD measurements were performed and the results are depicted Figure 4.1d-f. Since the different states have microstructures of very different length scales, data in Figure 4.14d-f were recorded at different magnifications. Within a sample, SEM images and corresponding EBSD mappings were taken at the same regions but not always at the same magnification. For the cold rolled material a smaller step size was chosen in order to resolve orientation changes within deformed grains. The EBSD results in Figure 4.1d-f represent crystal orientations of the material parallel to the sheet normal.

After the hot rolling process, a random grain morphology developed with a grain size that is smaller than what was found for the heat treated DC04 steel. The EBSD and SEM images of the cold rolled DC04 steel show a grain morphology that is elongated along the rolling direction. Compared to the hot rolled material, a large amount of undetected points (white pixels) are visible in the EBSD scan. These white pixels result from distorted diffraction patterns that cannot be identified by the EBSD software. Distorted patterns can for example occur due to multiple crystal orientations at one data acquisition point or due to smeared out diffraction patterns in highly deformed regions. The orientations in the cold rolled material (Figure 4.1e) vary within the grains which show that the grains were strongly deformed during the cold rolling process. Details on the misorientation within the grains will be given in section 4.2 and section 5.2. After the heat treatment process larger grains remain elongated along the rolling direction but this effect is less distinctive than in the cold rolled state. The prevalence of the blue color in the EBSD map shows that the grains have a preferred crystal orientation with the <111> directions parallel to the sheet normal.

The three dimensional crystal information can be visualized by plotting pole figures for the <111>, <110> and <001> orientations. All pole figures were calculated from the EBSD results in Figure 4.1. Figure 4.2 shows pole figures that were calculated for the DC04 material after each process step. The results for the hot rolled material show that the <111> and <001> crystal orientations are randomly distributed. The {110} pole figure shows a slightly higher intensity in the center of the pole figure which indicates that the grains have a preferred <110> orientation perpendicular to the sheet. The maximum intensity found for the hot rolled material is two or three times lower than what was found for the cold rolled and heat treated DC04 steel which implies that this orientation is only slightly preferred. After the cold rolling process, the DC04 steel has strong <111> and <001> fiber texture components. The intensity scale bar shows that the <111> fibers have a higher intensity than the <001> fibers. The <110> orientations are preferentially aligned along the rolling direction. After the heat treatment process, the maximum intensity in the center of the {001} pole figure is no longer present. Instead a {111} fiber texture developed with a maximum intensity that is three times higher than what was found for the cold rolled DC04 steel. The <110> directions that are aligned parallel to the rolling direction are still existing but are less distinctive than in the cold rolled material.

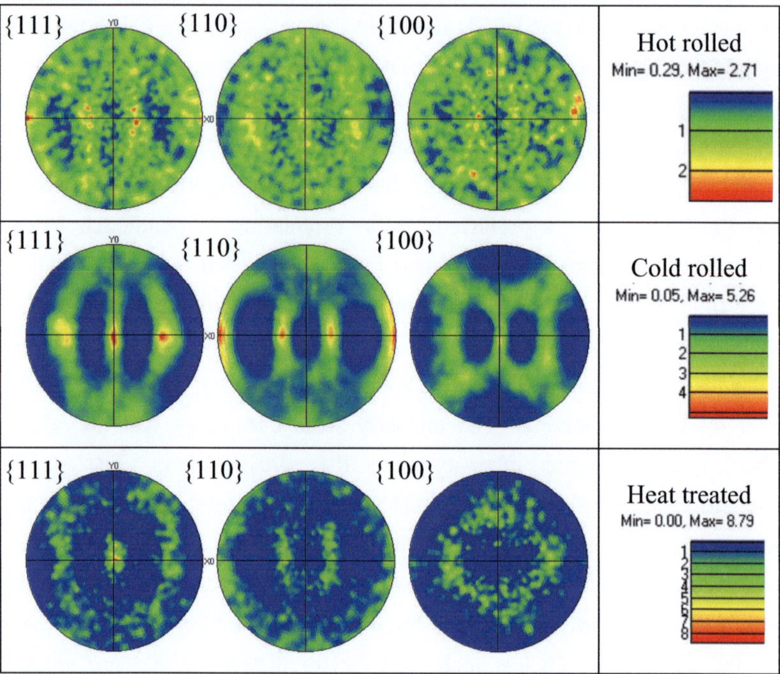

Figure 4.2: Calculated {111}, {100} and {110} pole figures for the hot rolled, cold rolled and heat treated DC04 steel. The pole figures were obtained from the EBSD results that are shown in Figure 4.1.

The maxima in the {110} and {111} pole figures of the cold rolled DC04 steel show that the material has pronounced α- and the γ-fibers. The γ-fiber consists of <111> orientations along the sheet normal and the α-fibers consist of <110> orientations along the rolling direction. These fibers form lines with higher intensity in the Euler space and are typically pronounced in low carbon steels (section 2.2). The pole figures do not contain complete three-dimensional orientation information and therefore cannot be used to analyze the distribution of the crystal orientations within these fibers. Figure 4.3 shows a projection of the reduced Euler space for the hot rolled, cold rolled and heat treated DC04 steel. In the hot rolled state, all Euler angles are almost randomly distributed in the Euler space but after the cold rolling and heat treatment process the Euler angles form regions of higher densities. A quantitative description of the distribution of the Euler angles can be obtained by calculating an orientation distribution function (ODF) (see Figure 4.3). The results in Figure 4.3 show that the hot rolled material has no clearly preferred crystal orientation. After the cold rolling and heat treatment process, the ODFs show higher densities in selected regions.

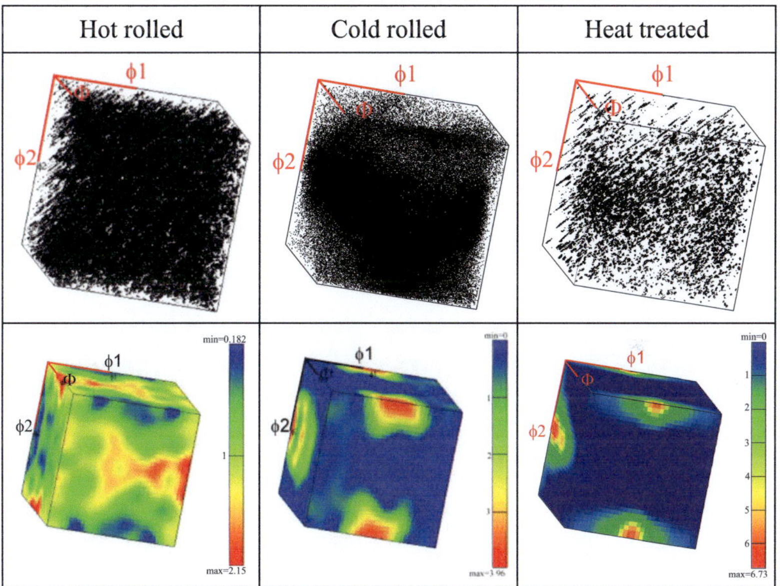

Figure 4.3: EBSD results of the hot rolled, cold rolled and heat treated DC04 steel that are plotted in the reduced Euler space and the calculated ODF.

In order to simplify the analysis of the calculated ODF, the Euler space is separated into equally spaced planes parallel to $\phi1$ (see Figure 4.4). The lines plotted in the ODF are contour lines that represent surfaces of the same density.

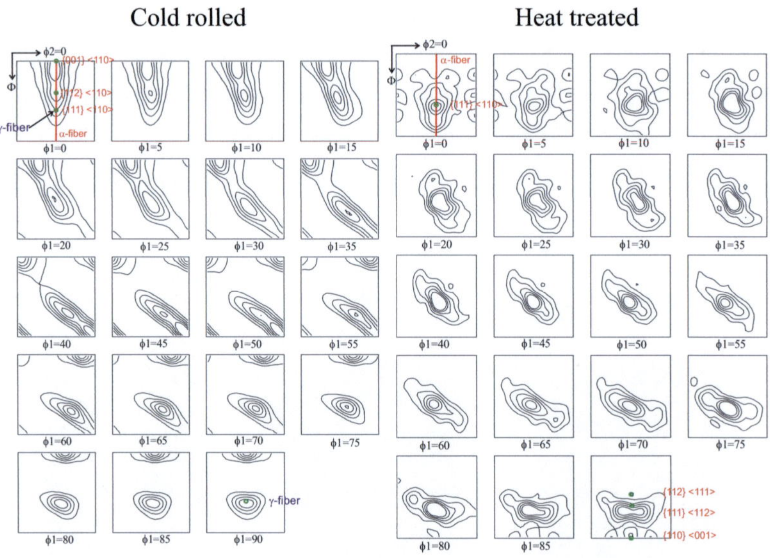

Figure 4.4: Calculated ODF for the cold rolled and heat treated DC04 steel. The α- and the γ-fiber are marked in red and green Additional green spots at $\phi1=90°$ show represent dominant orientations within the ε-fiber.

Figure 2.8 in chapter 2.2.2 shows the locations of the α- and γ-fibers in the Euler space. The α-fiber has the Euler angles $\phi1=0°$, $0°<\Phi<90°$ and $\phi2=45°$ and is highlighted as red line in the ODF (Figure 4.4) and the γ-fiber has Euler angles $0°<\phi1<90°$, $\Phi=54.7$ and $\phi2=45°$ and intersects with the α-fiber at $\Phi=54.7$. In the ODF, the γ-fiber is marked as green spots at $\phi1=0$ and $\phi1=90°$ although every cross section contains the γ-fiber at this location. On the α-fiber, important crystal orientations that are commonly found for bcc metals are marked in green (including the intersection between α and γ-fiber).

In order to investigate the dominant orientations in the α- and the γ-fiber, these two fibers are analyzed in detail. In Figure 2.7 (chapter 2.2.2) the locations of these two fibers within the Euler space are depicted. The figure shows that the α-fiber and the γ-fiber can be depicted by plotting the cross section parallel to $\phi1=0°$ and parallel to $\phi2=45°$ respectively. Figure 4.5a and b show sections that are plotted for $\phi1=0°$ that comprise the α-fiber and Figure 4.5c and d cross sections for $\phi2=45°$ that include the γ-fiber and parts of the α-fiber. The colors represent areas with high (red) and low (green) densities of orientations. After the cold rolling process, the maximum intensity for the α-fiber lies at $\Phi=0°$. After the heat treatment process this maximum is shifted to a higher Φ value that lies at the intersection of the α- and the γ-fiber and the intensity is reduced. Within the γ-fiber three maxima develop during the annealing process.

Figure 4.5a-d contain black lines that are drawn along the α-fiber and the γ-fibers of the cold rolled and heat treated DC04 steel. Figure 4.5e and f, shows ODF plots along these black lines as a function of the angles Φ for the α-fiber and as a function of $\phi1$ for the γ-fiber. After the heat treatment process, the maximum on the α-fiber is shifted to higher Φ values. On the γ-fiber (Figure 4.5f), the overall intensities increased after the heating and more structure in the plot became visible.

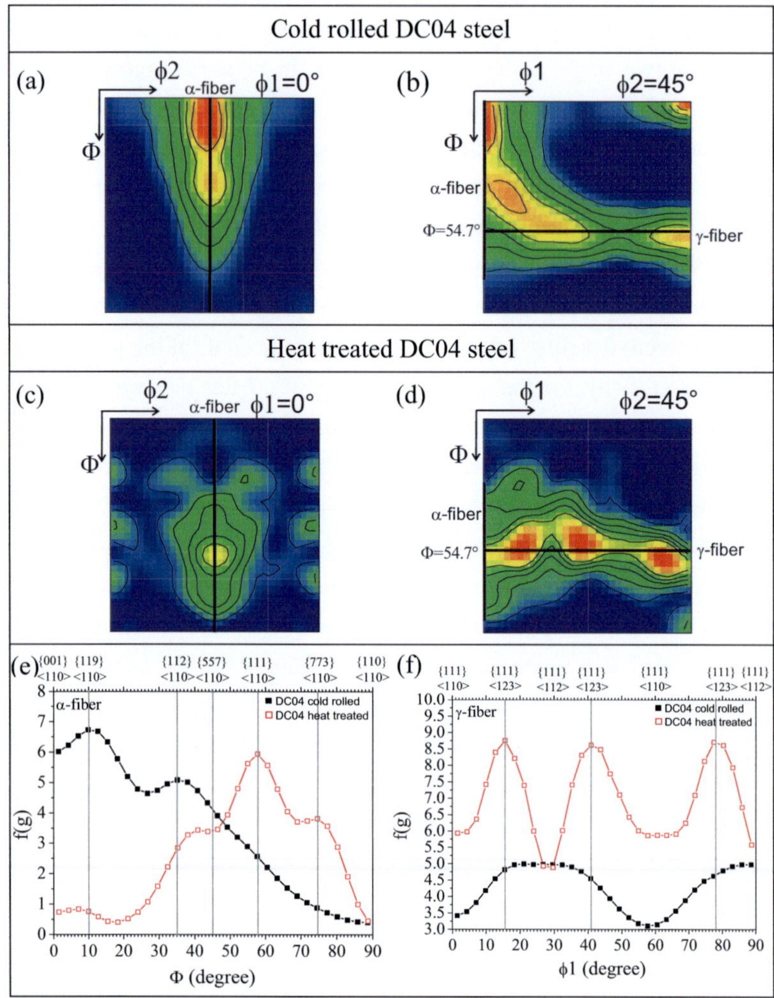

Figure 4.5: (a) – (d) Color coded cross sections of the Euler space for φ1=0 and Φ=45°which contain the α-fiber and the γ-fiber. Red describes a high density and green a low density of orientations at a given point in orientation space. (e)-(f) ODF plots that show the α-fibers and the γ-fibers of the cold rolled and heat treated DC04 steel

In the following the major orientations on the two fibers will be addressed by using their orientations relative to the sample normal (plane of the sheet) {uvw} and the rolling direction <hkl>. The α–fiber of the cold rolled state shows peaks at $\Phi=10°$ and $\Phi=35°$ which correspond to {119} <110> and {112} <110> orientations. After the heat treatment, the density of the {119} <110> orientations decreases whereas the {111} <110> orientations become dominant. The γ-fiber of the cold rolled material has a maximum at $\phi1=90°$ which corresponds to {111} <112> orientations. After annealing, the densities of the γ-fiber are shifted to higher values and shows pronounced peaks for {111} <123> orientations.

The development of the texture components during annealing of the cold rolled DC04 steel will be investigated in detail in the next chapter in which ODF after different heating steps are presented.

4.2 Texture evolution and grain growth during heating experiments of cold rolled DC04 steel

In order to investigate the evolution of the microstructure of DC04 steel during heat treatment, an experiment was performed where the temperature of a cold rolled steel sample was slowly increased up to 640°C within 10 heating steps listed in Table 4.1. The EBSD data were acquired together with Moritz Wenk and served as reference data for his diploma thesis that focused on the development of a software tool to analyze EBSD data [32]. Figure 4.6a-d show the calculated ODF plots for the α-fibers and the γ-fibers of the initial cold rolled material and after each individual heating step. The results are compared to data obtained from the heat treated DC04 steel as received from the manufacturer (section 4.1).

Slow heating procedure		
Heating steps	Temperature	Annealing time
Heating 1	594°C	7 min
Heating 2	596°C	5 min
Heating 3	598°C	18 min
Heating 4	610°C	15 min
Heating 5	631°C	10 min
Heating 6	652°C	18 min
Heating 7	647°C	20 min
Heating 8	647°C	20 min
Heating 9	647°C	60 min
Hetaing 10	647°C	720 min
Fast heating procedure		
Heating 1	679°C	33 min

Table 4.1: Heating experiments that were performed on the cold rolled DC04 steel

The first six heating steps were performed between 580-650°C. The α-fibers ($\phi1=0°$ and $\phi2=45°$) and γ-fibers ($\Phi=45.7°$ and $\phi2=45°$) are plotted in Figure 4.6a and b and the results are compared with the initial cold rolled material. The data show only little changes within the α–fibers and the γ–fibers during the first six heating steps. Due to the symmetry of the cubic crystal, the γ-fiber is only plotted from $60°<\phi1<90°$.

The heating steps 7-10 were all performed at 647°C and the results of the calculated ODFs are displayed in Figure 4.6c and d. After heating step 7 at 640°C first changes could be observed on the α-fiber. During further heating, the density of the {111} <110> orientations starts to increase and reach values that are higher than what was found for the heat treated DC04 steel. The densities of {001} <110> orientations continuously decrease after each heating step and reach almost the same value than in the heat treated material. The shoulder for the {773} <110> orientations that is found in the heat treated DC04 steel did not develop during these slow heat treatment experiments.

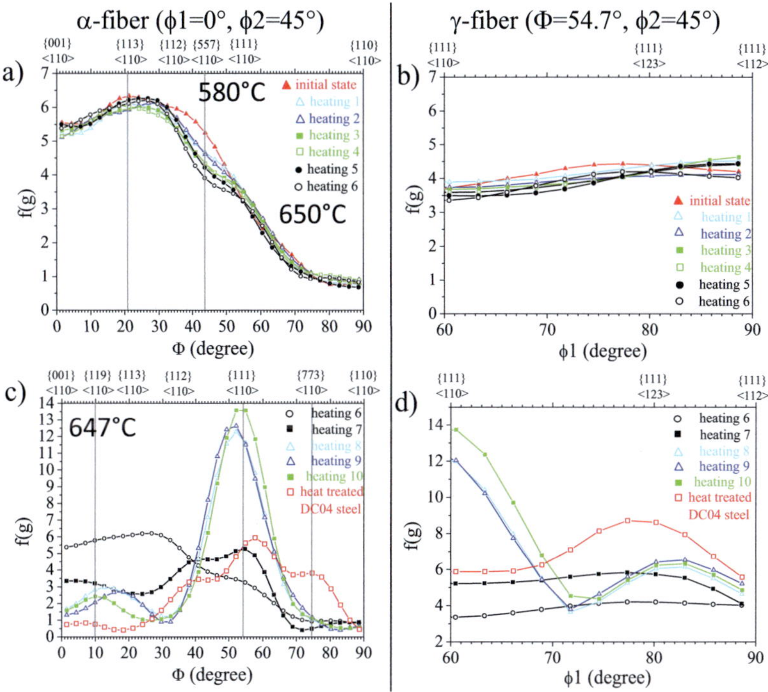

Figure 4.6: α-fibers and γ-fibers of the cold rolled DC04 steel and after the heating experiments. The results are compared to data obtained from the heat treated DC04 steel as received directly from the production process.

During the heat treatment at 640°C (heating step 7-10), the integrated density of the γ-fiber rises. This implies that the amount of <111> orientations parallel to the sheet normal increases. After this overall increase, individual orientations within the γ-fiber start to become prominent. Within the γ-fiber, the density of the <110> orientations increases and reaches values that are again more than twice as high as in the heat treated DC04 steel. The density of <112> orientations increases as well during the heating experiment and reaches almost the same value as found for the heat treated material. Additionally, density around $\phi1=83°$ develops which corresponds to the $\{111\}$ $<123>$ orientations. This peak was also observed in the heat treated material but slightly shifted and at a higher intensity. In order to assure statistical relevance, this texture data was collected on an area of 848 μm by 644 μm using a step size of 4 μm. To spatially observe the microstructure, additional EBSD maps were taken at each heating step.

EBSD measurements were also performed in smaller regions (318 μm by 241 μm) at a step size of 1.5 μm. Figure 4.7 shows the microstructure of the initial state and after heating step 1-10. In chapter 2.1.5, a concept for the estimation of the density of geometrically necessary dislocation (GNDs) with a rather simple model was introduced and is now applied (Figure 4.7). The EBSD results obtained from the initial cold rolled DC04 steel show that the grains are elongated along the rolling direction. The changes of the colors within the grains indicate that the material was strongly deformed during the rolling process and that a high density of dislocations is stored within the grains. The corresponding map shows a density on the order of 10^{14} /m^2 as a result of the cold rolling process. In the dislocation density map, regions can be observed where the dislocation density is lower with values in the 10^{13} /m^2 range which can be ascribed to grains with $<001>$ orientation parallel to the sheet normal. The map indicates that grains with $<111>$ orientations have a higher dislocation density than grains with $<001>$ orientation.

After the first heating step at 580°C, the GND density map shows regions where the dislocation density decreased to values of roughly 10^{12} /m^2. In the EBSD data, freshly formed grains can be identified corresponding to these regions. Some of the new grains show variations in the dislocation density. Not much changes after the following heating steps (heating 2-heating 5). After increasing the temperature to 647°C (heating 6), the dislocation density in the new grains became more uniform and the grains started to grow. Also the formation of new grains was observed. After heating 7 (20 minutes at 640°C), strong grain growth occurred. Although the grain growth rapidly advances, strong variations exist in the growing grains. After the last heating (heating 10), the grains that were already present after heating 7 grew further and the dislocation density within these grains reduced to values of roughly 10^{12} /m^2.

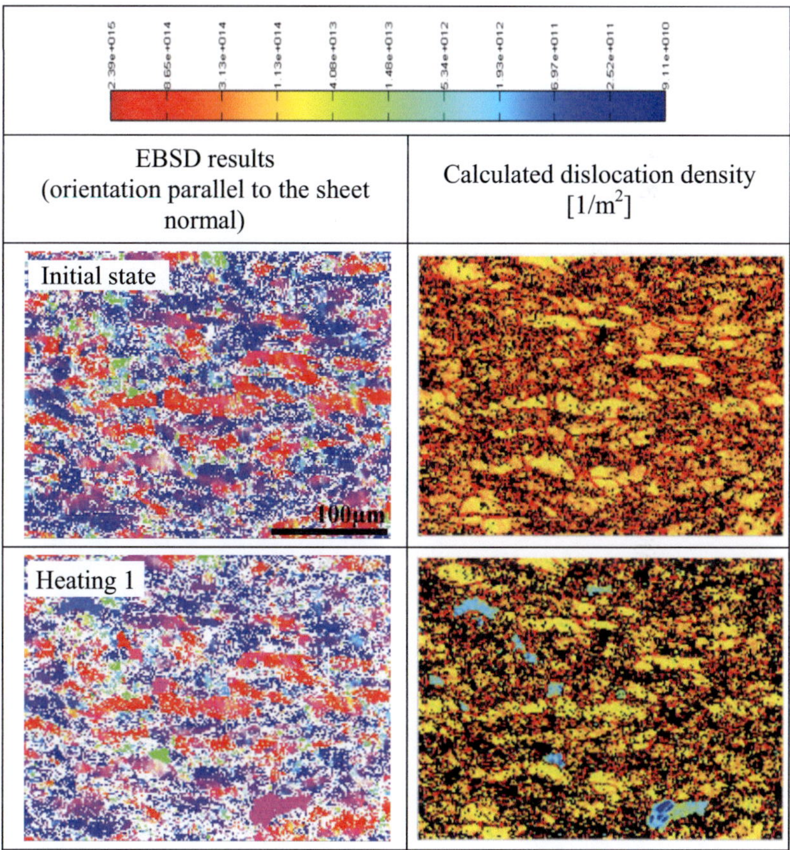

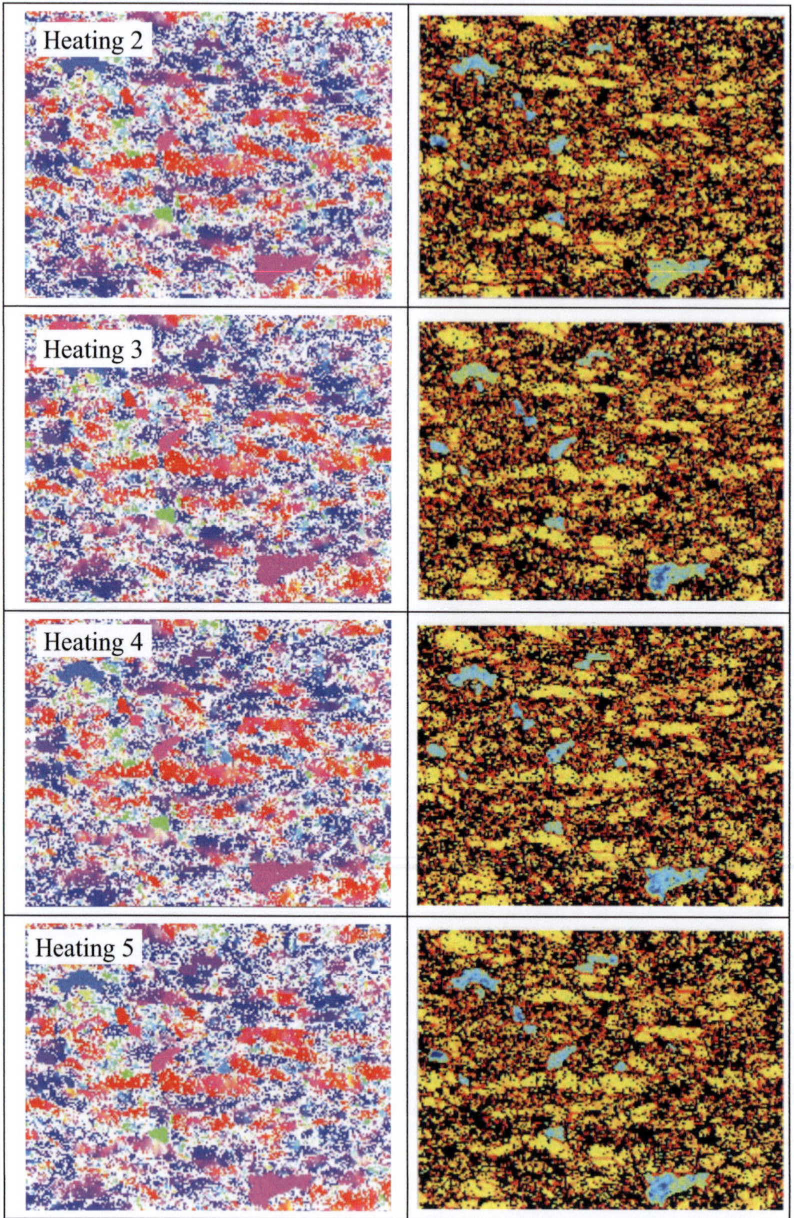
Heating 2
Heating 3
Heating 4
Heating 5

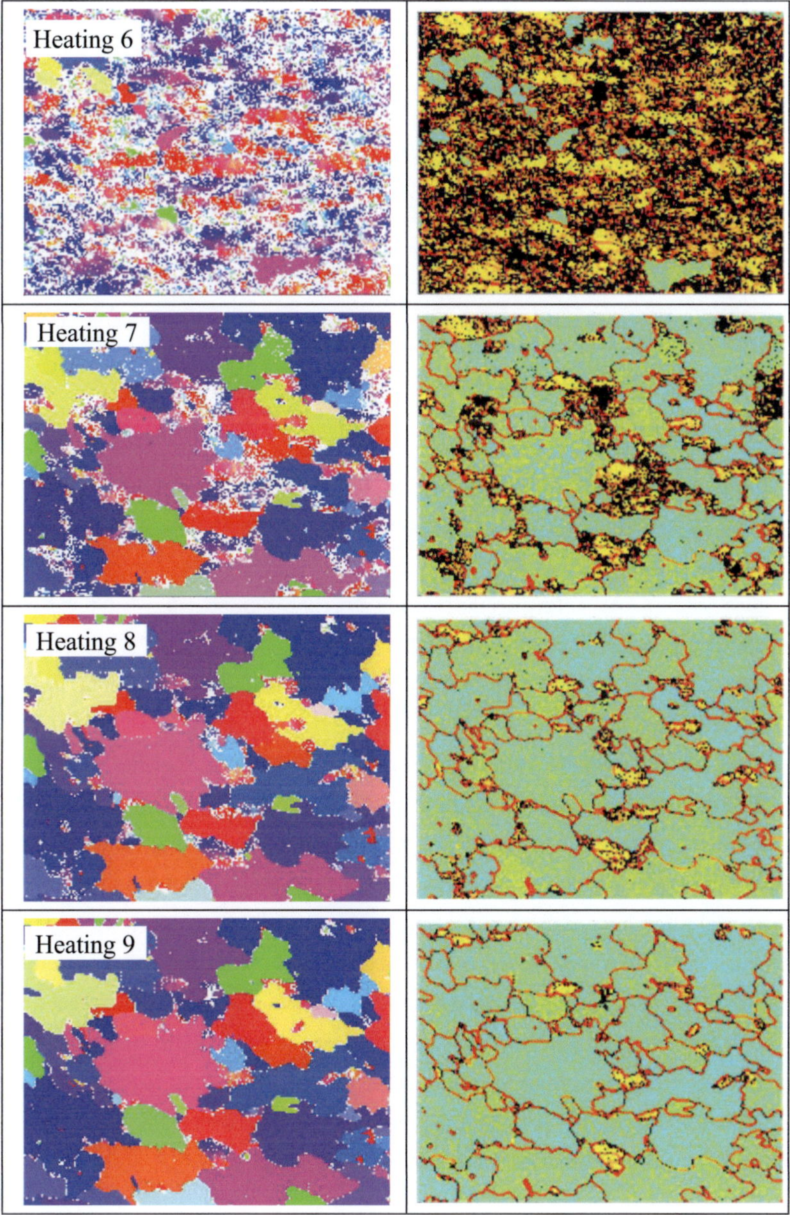

Heating 6
Heating 7
Heating 8
Heating 9

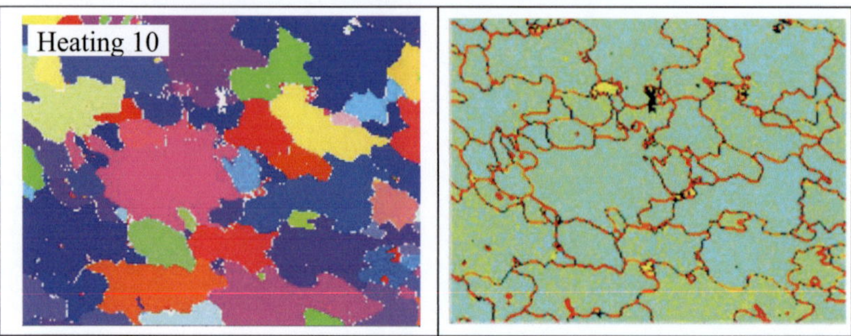

Figure 4.7: EBSD maps and corresponding GND densities of the cold rolled DC04 steel The presented results were received in cooperation with Moritz Wenk.

The data in Figure 4.7 allowed for the analysis of the nucleation and growth of individual grains. The inital state is compared to the state after first heating at 580°C and heating 6 at 640°C. In the EBSD map in Figure 4.8a (first heating), six regions are highlighted, in which freshly formed grains are observed. These regions are plotted with higher magnification in Figure 4.8b and are compared to the same regions in the cold rolled material.

The results indicate that grain 1 arose from a region that had a similar orientation in the initial cold rolled material. For the grains 2, 3, 5 and 6, small sites were already observed in the initial cold rolled material that have approximately the same orientations as the grains after heating 6. The observed sites are marked in the image. Grain 4 is an example of a grain whose nucleation sites could not be observed in the initial microstructure which may also be a consequence of the high amount of undetectable orientations in this region. Although only inverse pole figure maps relative to the sheet are used here to show grain growth, it was confirmed that the arguments about the nucleation sites still hold, when the full three-dimensional crystal orientation is taken into account [32].

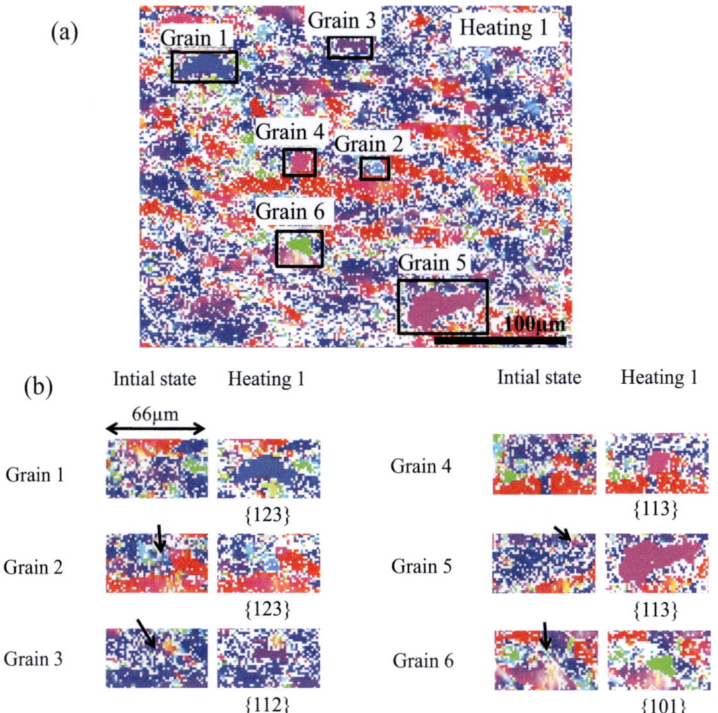

Figure 4.8: (a) shows the EBSD map after heating 1 in which freshly formed grains are marked (grain 1-6). These regions are depicted with higher magnification in (b) and the regions are compared the cold rolled DC04 steel.

After heating 6 the formation of more grains could be observed. The location of three freshly formed grains (grain 7-9) is marked in the EBSD map in Figure 4.9a. The same regions are plotted at higher magnification in Figure 4.9b and are compared to the microstructure of the initial cold rolled material and to the microstructure resulting from heating 6.

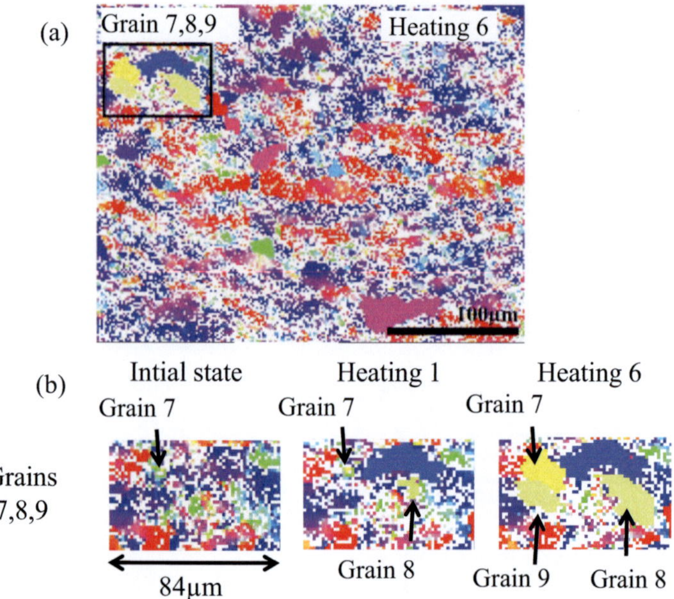

Figure 4.9: (a) shows the EBSD maps after heating 6 in which three freshly formed grains are highlighted (grains 7-9).

For grain 7 it seems that a very small grain or site with similar orientation was already present in the initial cold rolled material which became larger in the subsequent heat treatment steps. After heating 1 the formation of grain 8 could be observed. For this orientation no nucleus is visible in the initial microstructure. Grain 9 formed after heating 6. Also for this grain no suitable nucleation site is visible after heating 1. Between heating 6 and heating 10, grains 8 and 9 coalesce to form a larger grain (Figure 4.7).

4.3   Influence of heating conditions on the microstructure

In order to investigate the influence of higher heating rates and heating temperatures on the microstructure of the DC04 steel, a heating experiment was performed at 680°C for 33 minutes. Figure 4.10a shows the EBSD results that were obtained after this fast heating experiment. The result is compared to the microstructure that developed after the slow heating experiment at 647°C introduced in the last section (section 4.2, Figure 4.7). The microstructure after this slow heating experiment is depicted in Figure 4.10b. Figure 4.10c shows the microstructure of heat treated DC04 steel as extracted from the steel production process (section 4.1, Figure 4.1f).

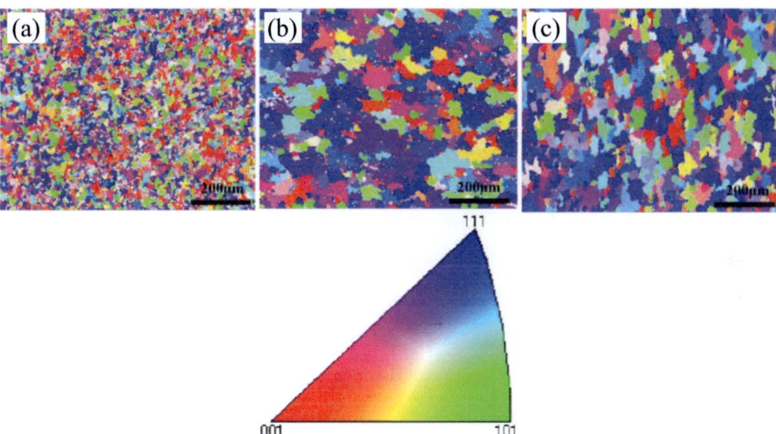

Figure 4.10: EBSD maps of DC04 steel that represent the crystal orientation parallel to the sheet normal after (a) a fast heating experiment at 680°C for 33 min, (b) a slow heating at 647°C (see section 4.2) and (c) of the heat treated DC04 steel as extracted from the steel manufacturing process (see section 4.1).

Among the three materials, the one with the fast heating to 680°C (Figure 4.10a) exhibits the smallest grain size. The distribution of color in this map also indicates that there is no or only little texture in this material. After the slow heating experiment (Figure 4.10b), the grains have a larger grain size than the material that was extracted from the industrial process chain (Figure 4.10c). In both materials a preferred <111> crystal orientation perpendicular to the sheet can be observed and in both cases the large grains are elongated along the rolling direction. Details on the texture can be observed in the pole figures in Figure 4.11.

Figure 4.11a shows pole figures after the fast heating process. The microstructure of this material has a rather random distribution of the <111>, <110> and <001> orientations. Despite this randomness, the {111} and {001} pole figures show slightly higher intensities in the center which means that the {111} and {001} planes are preferentially aligned parallel to the sheet. The {110} pole figure shows a maximum along the rolling direction. In general, the maximum intensities for all three pole figures are three times lower than what was found after the slow heating experiment (Figure 4.11b) or for the heat treated material from the steel production process (Figure 4.11c). After the slow heating process (Figure 4.11b), the material exhibits a similar pole figure than the DC04 steel from production (Figure 4.11c). In both materials, α- and γ-fiber components can be observed. The γ-fiber corresponds to the center pole in the {111} pole figure. Evidence for the α-fiber is visbile in the {110} pole figures from the symmetry of this figure. Ideally poles appear at the perimeter as in the cold rolled material in Figure 4.2b, but here the intensity of this fiber is rather weak and only the rotational symmetry along the <110> direction indicates the presence of the α-fiber in the {110} pole figure. After the slow heating experiment the orientations seem to be less discrete and more clustered which is likely an effect of the larger grains. The intensities of the poles are higher than what was found for the DC04 steel as received from the production process.

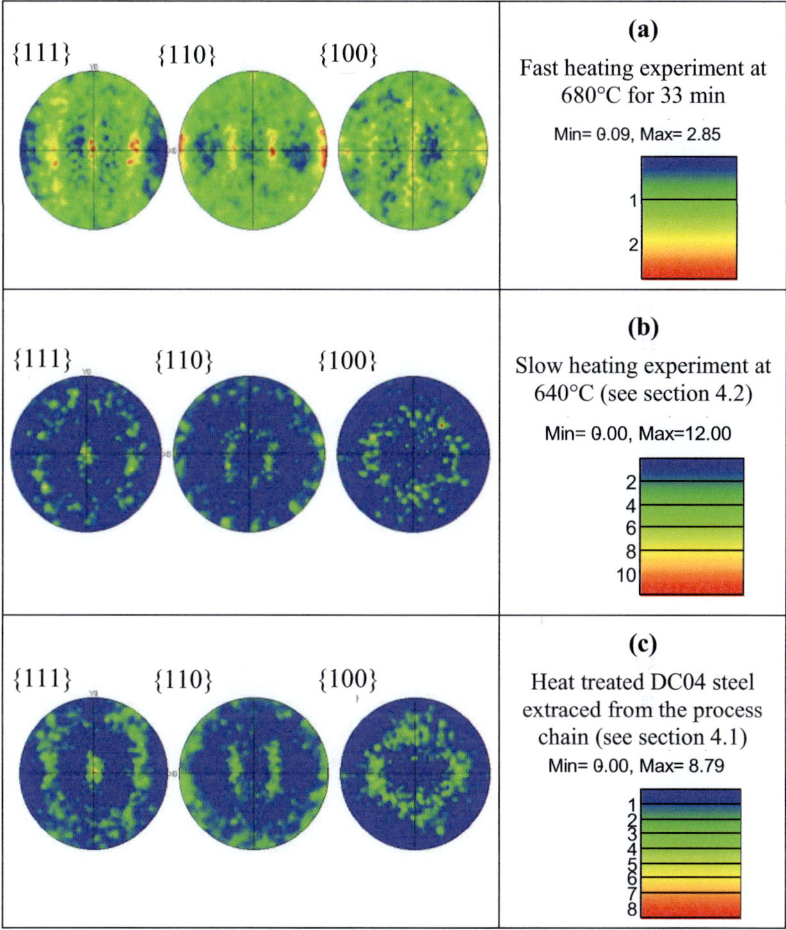

Figure 4.11: Pole figures measured after the three different heating processes applied to cold rolled DC04 steel. The pole figures were generated from the EBSD data in Figure 4.10.

For detailed analysis, the ODF plots that correspond to the α-fiber and the γ-fibers of the DC04 steels after the three different heating processes are plotted in Figure 4.12. In addition, the fibers of the cold rolled DC04 steel are depicted. After the fast heating experiment to 680°C for 33 min, the material shows a weak minimum in the α-fiber for the {111} <110> orientations and slightly higher intensities for the {119} <110> orientations. In both, the α-fiber and the γ-fiber, the overall intensity is lowered which indicates that orientation information is lost and that a transition to a more random orientation occurs during the fast heating experiment.

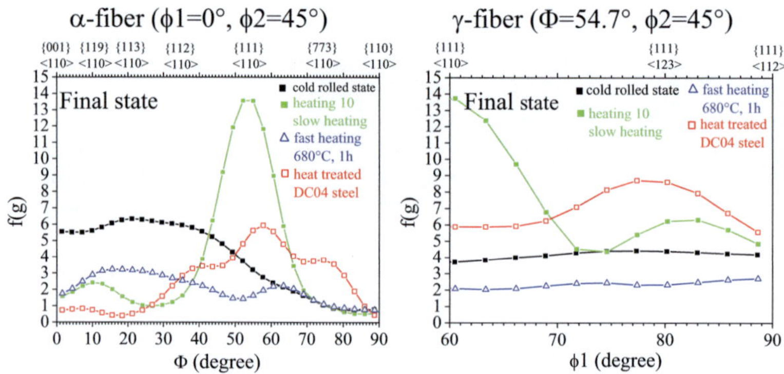

Figure 4.12: Calculated ODF of the α- and γ-fiber for the cold rolled DC04 steel, the heat treated material received from the process chain after two different heating procedures

Figure 4.13 shows the EBSD maps of the cold rolled material and after the fast heating experiment to 680°C. The orientations are plotted along the rolling direction Figure 4.13a, b and parallel to the sheet normal Figure 4.13c, d. The EBSD maps in Figure 4.13a, b show, that the amount of <110> and <111> orientations do not significantly change during the heating experiments. Parallel to the sheet normal, the amount of <110> orientations increases and the amount of <111> orientations decreases.

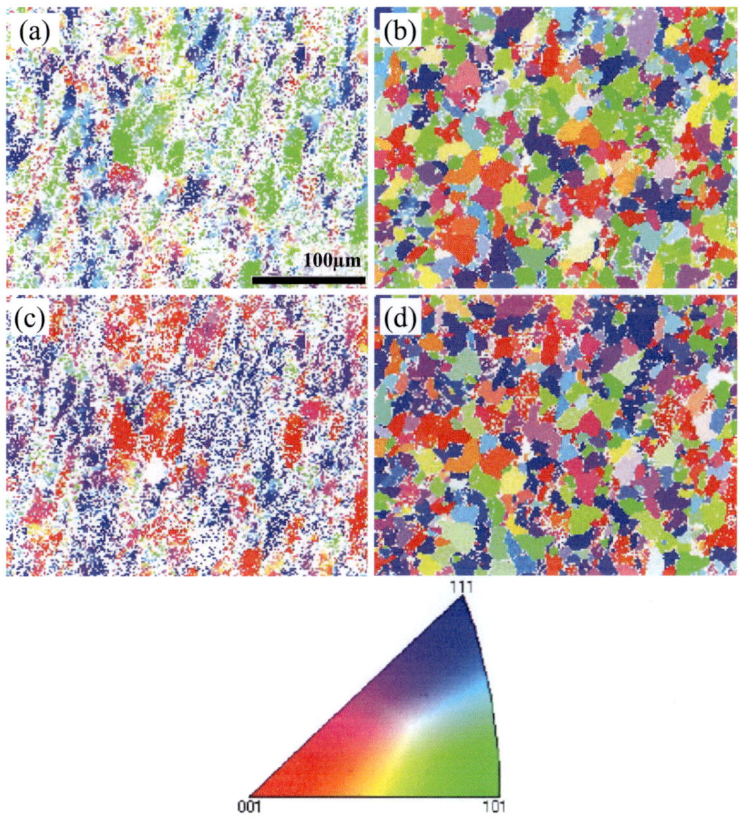

Figure 4.13: EBSD maps of (a), (c) the cold rolled DC04 steel and (b), (d) after a heating experiment to 680°C for 33 min. The EBSD maps in (a) and (b) present the crystal orientations parallel to the rolling direction and in (c) and (d) the out of plane orientations.

4.4 Discussion

In the previous sections of this chapter the results from microstructure analysis after cold rolling and after different heating experiments were presented. In this section, the changes in the microstructure and in particular the evolution of the different texture components under different heating conditions will be discussed. Furthermore, this section focuses on grain growth and recrystallization and how these processes are affected by different heating conditions.

4.4.1 Texture of cold rolled DC04 steel

The ODF plots that were presented for the cold rolled DC04 steel (Figure 4.4) show similarities to the ODF plots that were determined by Von Schlippenbach [27] for a cold rolled AI steel with similar chemical composition (section 2.2.2, Figure 2.11). Differences can be found in the location of the maxima in both ODFs. Therefore, the corresponding α- and γ-fibers are compared. The α- and γ-fibers that were determined by Von Schlippenbach (Figure 2.12) have maxima at {111} <110>. The fibers that were determined for the cold rolled DC04 steel that have their maxima at {001} <110> and {111} <112> (Figure 4.5e, f). A possible reason may be that the ODF plots from the literature were determined after a rolling reduction of 80 %. Raabe et al. [23] determined the evolution of the α- and γ-fibers after different rolling reductions of an IF steel (section 2.2.2, Figure 2.10). After a rolling reduction of ~70 % the α-fiber have maxima for {001} <110> and {112} <110> orientations although the intensities are different compared to the cold rolled DC04 steel. The γ-fiber plot that was determined by Raabe et al. also shows a maximum for the {111} <112> orientations.

4.4.2 Texture evolution during slow heating of cold rolled DC04 steel

The texture that forms during the slow heating experiment is not random and is still governed by a γ-fiber and to some extent by the α-fiber. This suggest that the in plane orientation that develops during the heat treatment is affected by the initial cold rolling. In the following, details on the texture evolution and mechanisms that can possibly lead to the observed textures will be mentioned.

Only minor changes occur until heating step 6 (Figure 4.6). Starting after heating 7, the orientation density in the γ-fiber plot isotropically increases (increase in <111> out of plane orientations). In parallel, the α-fiber also changes and preferred orientations already become visible. This change in the α-fiber consists of a decrease in the density between {001} and {112} and a concurrent increase of density at {111}. In general, two mechanisms come into question that may lead to an increase in the density of the <111> out of plane orientations. First, new {111} grains may form within the deformed {001} and {112} grains which would also decrease the density of these two orientations. Second, existing {111} grains may grow in excess of grains with other orientations. Considering the microstructure in Figure 4.7, it can be seen that during heating 7 dramatic changes happen and afterwards a large area fraction is covered by a recrystallized region. These changes happened after only 20 minutes at 647°C. During this short time very large grains appeared. A closer look at the large grains indicates that {111} grains with high levels of misorientation grew in excess of their neighbors. This suggests that the second mechanism for the increase of {111} density is responsible and that the highly deformed regions that are visible after heating 6 (Figure 4.7) are afterwards preferentially covered by new {111} grains. The increase of the {111} grains during heat treatment of cold rolled steel was also observed in the literature [85, 86]. The increase of the {111} density also seems to appear in Figure 4.7 where the fraction of blue grains has increased after heating 7, in accordance to the isotropic increase in the γ-fiber plot. It should be noted here that the texture data (Figure 4.6) was recorded in regions that are much larger than the regions that display the microstructure (Figure 4.7). Nevertheless the smaller regions are used here to illustrate the spatial distribution of the grains. In the initial microstructure (Figure 4.7), {111} grains generally have a higher dislocation density than {001} grains (Figure 4.7). This is also observed in the literature on cold rolled α-Fe [87]. A possible reason for the preferred growth of {111} grains may be due to the fact that the high density of dislocations in the {111} regions offers stored plastic energy that can be released during recovery and grain growth. The assumption is confirmed by Figure 2.13 in section 2.2.3 which shows that the reduction of the plastically stored energy is the main driving force at high dislocation densities. Both processes may not necessarily lead to changes in local orientation.

Strong changes in the γ- and α-fibers can be detected after heating 8. Although heating 8 occurred at 648°C for 20 min, with parameters that are very close to heating 7 a very different effect in the texture is achieved. At first glance (Figure 4.7) not much changes in the microstructure, but from Figure 4.6 it can be seen that the texture strongly changes. Here preferred orientations form along the γ-fiber which are the {111} <011> orientations and orientations not far from {111} <123>. Also the already existing {111} <110> orientations in the α-fiber plot dramatically strengthens. Combining the information from both fibers shows that the area fraction of {111} <110> orientations, that belong to both fibers increases strongly. From this point on only little changes occurred in the texture and upon heating. This is even valid after heating 10 that took 24 h at the same temperature. The microstructure in Figure 4.7 suggests that the change in the texture is due to the second mechanism in which grains with the respective orientation grow to increase the orientation density. After heating 7, regions with high dislocation density remained. A possible driving force for the growth of grains during heating 7 and the associated change in texture may be the reduction in stored plastic energy. The fact that this leads to the observed very strong sharpening of the texture may be partly due to the fact that many of the remaining regions are already <111> oriented out of plane. Nevertheless it is surprising that also a strong prevalence of the <110> orientations along the rolling direction forms. This seems to be due to the cold rolling step since the {111} <110> orientation lies at the intersection between α-fiber and γ-fiber which are both common for cold rolled bcc metals. Altogether a distinct sequential evolution of the microstructure was observed. First a strong effect in the microstructure (spatial changes) and subsequently a strong effect in the texture (orientation changes) occurred.

4.4.3   Grain nucleation and growth

In the experiments, EBSD maps were taken after all heating steps. This allowed the investigation of the development of the microstructure. By comparing EBSD maps (Figure 4.8) before and after heating 1 (594°C for 7min), small changes in the microstructure were detected. In most cases it seems that the orientations of the new grains were already present in the cold rolled material. The formation of subgrains that can occur during recovery (section 2.2.3) was not confirmed in our studies. These observations lead to the assumption that nucleation processes do not contribute to recrystallization in DC04 steel. From the few observations that were made no preferred crystal orientation can be identified and it seems that these early steps of grain growth happen for random orientations. In most cases the new grains were located in previously highly deformed regions. In a few cases the orientation of the growing grain could not be found in the deformed microstructure but it may be the case that this is only due to the poor quality of the diffraction patterns in the deformed regions. The very small nuclei that existed beforehand seem to grow into deformed regions. It is plausible that this early grain growth process is driven by a reduction in stored plastic energy.

During the course of growth over several heating steps (Figure 4.7), the density of dislocations changes in the growing grains. Also the appearance and disappearance of dislocation networks was observed. These observations demonstrate that the new grains are not fully relaxed and that mechanical processes can even strongly affect a growing grain. Further indications for the fact that high mechanical stresses exist during grain growth were found by a calculation of mechanical stresses from the diffraction patterns that were recorded in the experiments presented here. Details on the stresses of growing grains can be found in [32].

The final grain size that was achieved in the slow heating experiment is larger than the grain size of the commercial material (heat treated material). This is not surprising since the heat treatment during steel production may occur at different temperatures. In the production process the heating phase is most likely significantly shorter. In order to assess how the heating time affects the texture and microstructure, the measurements were complemented by a fast heating cycle.

4.4.4 Influence of different heating conditions on the microstructure

In the fast heating experiment the temperature of the sample was kept at 680°C for 33 minutes. The resulting microstructure in Figure 4.10 is composed of small grains with a rather uniform shape. It strongly differs from the microstructure that developed after the slow heating experiment and also from the microstructure of the material that was extracted from the production process. Although the pole figures (Figure 4.11a) show α-fiber and γ-fiber components for the fast heated sample, the orientations are more randomly distributed than in the other two cases. This is confirmed by the ODF in Figure 4.12 where the overall intensity for this material is always lowest.

Similarities can be found between the pole figure of the DC04 steel after the fast heating experiment and the pole figure of the cold rolled DC04 steel in Figure 4.2. The pole figures of both materials have maxima at the <111> and <001> orientations parallel to the sheet normal and maxima at the <011> orientations parallel to the rolling direction. In contrast to the cold rolled DC04 steel, the maxima that developed after heating at 680°C are less pronounced. From these observations it may be speculated that the recrystallization process was not fully completed after the 33 minutes. After the fast heating experiment the microstructure also differs significantly from what was observed during the slow heating experiments (between heating 1 and 10) in Figure 4.7. Two possibilities may be considered to explain the observations. Either the number of nuclei for grain growth may be different at the two different temperatures or there may be competing mechanisms during microstructure evolution.

The first case follows the classic picture of microstructure evolution. There recovery takes place at relatively low temperatures before the recrystallization process becomes active (section 2.2.3). In this case, the recovery process defines the distribution of the nucleation sites for grain growth and subsequently at higher temperature the grains start to grow. This leads to different microstructures depending what has happened before during recovery. Humphrey and Heatherly [40] describe that the recrystallization temperature decreases with increasing stored plastic deformation. At lower heating temperatures, the recrystallization process is therefore limited to regions that contain sufficiently high dislocation densities. Gottstein [31] describes that the number of available nuclei decreases with decreasing heating rate due to recovery processes that occur during slow heating processes. The combination of both effects may lead to the observed microstructures. Extended recovery at lower temperatures leads to a low density of nuclei. At higher heating temperature and high heating rates, the number of available nuclei will increase and a microstructure with smaller grains is expected to form during grain growth.

The initial cold rolled microstructure (Figure 4.7) is inhomogeneous and contains regions with higher and lower dislocation densities. During the slow heating experiment the first freshly formed grains were observed in regions that have high dislocation densities. After heating 7, {111} grains that also contain high densities of dislocations started to grow in excess of the neighboring grains. This shows, that only a few selected sites started to grow at lower temperatures and formed a microstructure with large grains. At higher heating temperatures it may be possible that regions with lower dislocation densities can additionally start to grow. This increasing number of growing regions would form a microstructure with smaller grains which could be observed after the fast heating experiment. This mechanism is typically observed after the primary recrystallization process of deformed materials at higher heating temperatures.

According to this picture it is surprising that almost no subgrain formation was found and that the microstructure evolved around existing orientations. Also surprising is the fact that there was almost no grain growth observed at the higher annealing temperature and that the microstructure did not change dramatically at the higher temperature. Alternatively to the classic description of microstructure evolution, a mechanism is discussed in the following where it will be assumed that the dominant mechanisms change when the sample is heated at different rate or to different temperatures.

In general, recovery and recrystallization processes control the evolution of the microstructure during heating experiments. In the slow heating experiments (section 4.2) the evolution of small nuclei was observed. It became clear that the microstructure evolution in this case is dominated by grain growth and that nuclei were most likely already present at the beginning. Therefore, for cold rolled DC04 the recrystallization process seems to be dominated by grain growth and less by grain nucleation. During grain growth, grain boundaries move. For large angle boundaries in pure metals this process is controlled by grain boundary diffusion [28]. Diffusion processes are generally described by the Arrhenius equation and depend on the temperature T, the activation energies Q, the diffusivities D_0 and the Boltzmann constant k.

$$D = D_0 \cdot exp\left[-\frac{Q}{kT}\right] \qquad (4.1)$$

Another mechanism that releases stored plastic energy is recovery, where stored dislocations and vacancies annihilate without changing the microstructure. For this to occur, dislocations need to glide or climb. In particular, the climb processes are rate limiting since they require volume diffusion. The diffusivity also follows equation (4.1). Diffusion of atoms inside the grain volume need a significantly higher activation energy Q_V than diffusion processes inside the grain boundaries Q_{GB}. In the literature [88, 89] activation energies of $Q_V = 2.48eV$ for the volume self-diffusion and $Q_{GB} = 1.81eV$ for grain boundary diffusion in α-Fe were reported. The pre-exponential factors have values of $D_{0_{GB}}$=1,27$\cdot$10^{-5} m^2/s [88] and D_{0_V}= 2.76$\cdot$10^{-4} m^2/s [89, 90] assuming a grain boundary width of 1 b. With these values the diffusion coefficients at 647°C and 680°C can be calculated (Table 4.2).

	640°C	680°C
Grain boundary diffusion	D_{GB}=7.63^{-13} m^2/s	D_{GB}=1.35$\cdot$10^{-12} m^2/s
Volume diffusion	D_V=9.88$\cdot$10^{-21} m^2/s	D_V=3.67$\cdot$10^{-20} m^2/s
D_{GB}/ D_V	77$\cdot$10^6	37$\cdot$10^6

Table 4.2: Diffusion rates calculated for α-Fe at 640°C and 680°C [88-90]

The microstructural evolution of the DC04 steel is a complex process based on mechanisms that are most likely dependent on individual grain boundaries on impurities and many other factors. Despite this complexity, in the following a very simple picture will be developed that tries to explain the different microstructures by a competition between recovery and recrystallization (Figure 4.14). This simple consideration can explain the differences and is in qualitative agreement with the observations made at both temperatures. At temperatures of 647°C, volume diffusion may still be too slow to lead to significant recovery and grain boundary diffusion may be the dominating mechanism that causes the reduction in stored plastic energy. In this case large grains form and changes in the texture of the cold rolled metal will occur. At 680°C, the situation may be very different. Here the diffusivity through the volume is roughly four times higher than at the lower temperature and dislocations or other defects in the grains may annihilate before grain boundary motion can begin. In this case the stored defect energy is already consumed and the driving force for subsequent grain growth is lacking. Therefore, no dramatic changes in the microstructure are expected and in particular the initial grain size will be conserved. This picture roughly describes the experimental observations, where smaller grains were found at the higher annealing temperature. In the experiment a temperature difference of only 33°C leads to this very marked difference in the microstructure. In order to further describe the simple concept of the competition of mechanisms a simple Gedankenexperiment will be performed in the following.

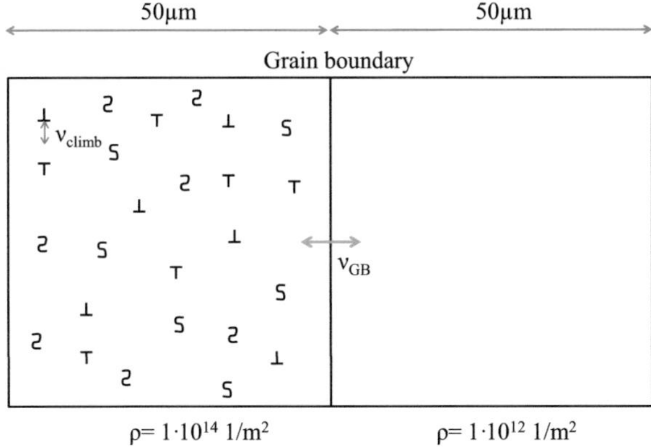

Figure 4.14: Schematic of two neighboring grains with a high (left) and a low (right) density of dislocations ρ. The average distance between the dislocations in the left grain are 50 nm. v_{climb} is the climb velocity of the dislocations and v_{GB} the velocity of the grain boundary that is moving into the grain with high dislocation density.

Figure 4.14 shows two neighboring grains. The grain on the left has a density of dislocations similar to the cold rolled material. The grain on the right has a lower dislocation density which is characteristic for the annealed material. The free energy of the material can be reduced by the removal of dislocations. This can happen by the two pathways already described. In the first case the grain boundary sweeps across the left grain. This process is diffusion limited and in order to move a high angle grain boundary, atoms need to diffuse across and along the boundary. The drift velocity v_{GB} of matter during diffusion is given by

$$v_{GB} = M\Delta G \qquad (4.2)$$

where M is a mobility and ΔG a driving force. Here ΔG is the decrease in free energy per unit volume when the atom diffuses from one side of the boundary to the other side. For the mobility of the boundary a simple model which is based on the Nernst-Einstein relation for high angle boundaries is used [91, 92].

$$M = \frac{D_{GB}\, b^2}{kT} \tag{4.3}$$

Assuming a boundary width of one Burgers vector b and an atomic volume Ω. The stored plastic energy of the left grain in Figure 4.14 is considered to be the driving force for atoms moving to the right grain. The energy per length of a dislocation line is given by $E/L \approx Gb^2/2$ with G being the shear moduls and L the length of the dislocation line. For a given dislocation density ρ the energy density of the plastically stored energy is $E/V = \rho Gb^2/2$. Since not all dislocations meet the interface a factor of 1/3 is introduced.

$$\Delta G = \frac{1}{3}\frac{\rho Gb^2}{2} \tag{4.4}$$

Equations (4.2), (4.3) and (4.5) yield

$$v_{GB} = \frac{D_{GB}\, b^4 \rho G}{6kT} \tag{4.5}$$

Inserting values into this equation will yield estimates for the velocity of the boundary during its motion into the plastically deformed region (right part of Figure 4.14). The mobility used before does not take into account microscopic mechanisms and is known to often lead to too high velocities for the interfaces [91].

In the second pathway, recovery takes place and dislocations can annihilate without causing strong microstructural changes. This may happen by several mechanisms but quite likely dislocation climb will be required. This is a slow process that is based on vacancy diffusion in the bulk. During climb, vacancies diffuse from or to the dislocation and cause its motion by growth or shrinkage of the inserted extra half plane representing the dislocation. The driving force for this process is more localized and is due to the mechanical stresses around the climbing dislocation. The climb velocity can be estimated [93] by using a cutoff radius of 1000b, which leads to the following relation for the velocity of climbing dislocations

$$v_{climb} \approx \frac{D_V \Omega}{kTb} \sigma \approx \frac{\sigma D_V b^2}{kT} \tag{4.6}$$

Using climb, dislocations can overcome obstacles. If there are no obstacles or only small obstacles around a dislocation, they will most likely move by glide or thermally activated glide. In order to estimate the velocity of recovery, only the most time consuming and therefore rate limiting step of dislocation climb is considered here. Although climb may require most of the time during the recovery process, it is most likely not the mechanism that will move dislocations by the largest distances. At this point a somewhat arbitrary assumption is made that only 10 % of the initial dislocation distance needs to be covered by climb. For a dislocation density of $\rho = 10^{14} \frac{1}{m^2}$, an average dislocation spacing of 100 nm results. If both dislocations approach each other, then only 5 nm need to be covered by climb. In the following the velocity and times for recovery/grain growth for the two pathways are compared at the two different temperatures. Table 4.3 shows the parameters that were used and Table 4.4 shows the results.

Shear Modulus of Fe	G=82.2 GPa [94]
Boltzmann konstant	k=1.381·10^{-23} J/K
Burgers vector of Fe	b=2.48·10^{-10} m
Avogadro Constant	N_A =6.022·10^{23} mol^{-1}
Lattice konstant of Fe	a=2.86·10^{-10} m
Stresses in the grains	σ=400 MPa [32]
Distance to climb	d=($1/\sqrt{\rho}$)

Table 4.3: Material parameters of Fe that were used to calculated the climb velocities of dislocations and the grain boundary velocity at 647°C and 680°C

	Grain growth		Recovery	
	V_{GB}(nm/s)	$t(s)$	V_{climb} (nm/s)	$t(s)$
T1 (647°C)	310	161	0,02	250
T2 (680°C)	530	94	0,07	71

Table 4.4: Estimated values for the velocities of the grain boundaries and for dislocation climb at 647°C and 680°C. The times are needed to form a grain with a grain size of 50μm by the respective process.

These values together with diffusivities from Table 4.2 give the times in Table 4.4 that are required to transform a grain (Figure 4.14) with a diameter (length) of 50 μm. From the times in Table 4.4 it can be seen that the controlling mechanism for microstructure evolution switches from grain growth at the lower temperature to recovery at the higher temperature. It has to be noted that the assumptions about the mechanisms that were made here are rather coarse and do not realistically reflect the underlying physical processes in detail. Even when just looking at the two selected processes, the definition of the structure independent mobility of grain boundaries and the criterion for the distance where annealing can take place appear to be very critical. It is therefore possible that there are errors in the velocities as large as one order of magnitude. Despite its simplicity this simple model demonstrates two of the very important mechanisms and illustrates the competition between recovery and recrystallization. It has to be noted that both mechanisms that were used here are not fully independent. During microstructure evolution they are expected to run in parallel so that they would compete. This competition would favor the dislocation climb process, because reductions in the dislocation density within one grain would reduce the driving force for grain growth (equation (4.6)) and therefore would slow down this process. The simple Gedankenexperiment illustrates typical dimensions and velocities of the two processes. The observed microstructure shows that the grain size and grain orientations do not significantly change after the fast heating process. This fortifies the assumption that the competition between recovery and grain growth is responsible for the strong temperature sensitivity of the microstructure of heat treated DC04 steel.

5. Size Dependent Micromechanical Behavior of DC04 Steel

This chapter focuses on the deformation behavior of DC04 steel on different length scales. From many fcc and bcc metals it is known that they exhibit a strengthening that is dependent on the sample size. In this work, the deformation behavior of the heat treated and cold rolled DC04 steel were investigated by microcompression experiments on single and polycrystalline pillars with the focus on sample size and microstructure. Here the influence of alloying elements, precipitates, different crystal orientations and predeformation on the size dependent deformation behavior of DC04 steel will be investigated. The results will be compared to the size dependent deformation behavior of single crystalline α-Fe which serves as a reference material since it has the same crystal structure as DC04 steel. The α-Fe data that is used here was measured by Daniel Kaufmann in his Ph.D. thesis [56].

5.1 Heat treated DC04 steel

The DC04 steel samples were taken during the steel production process right after a heat treatment step. Details of this heat treatment are not known but the resulting microstructure showed an average grain sizes in the range of 50 µm (see Figure 4.1). The pillars were machined into different grains and have <111>, <123> and <011> orientations along their axis. Figure 5.1 shows examples of two single crystalline DC04 steel pillars with diameter of 2 µm and 4 µm that were micro machined in the heat treated material by the top down milling method.

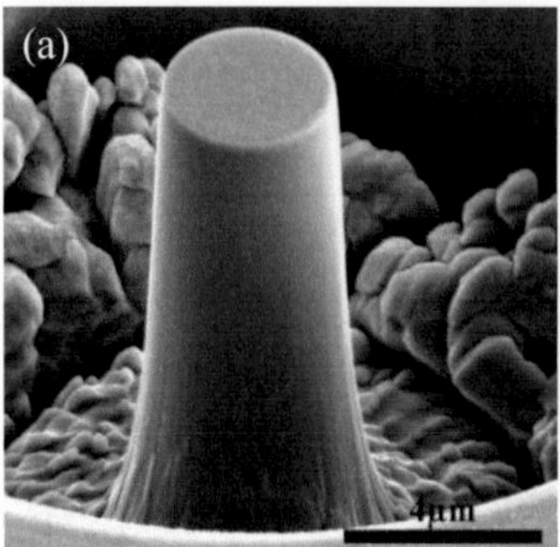
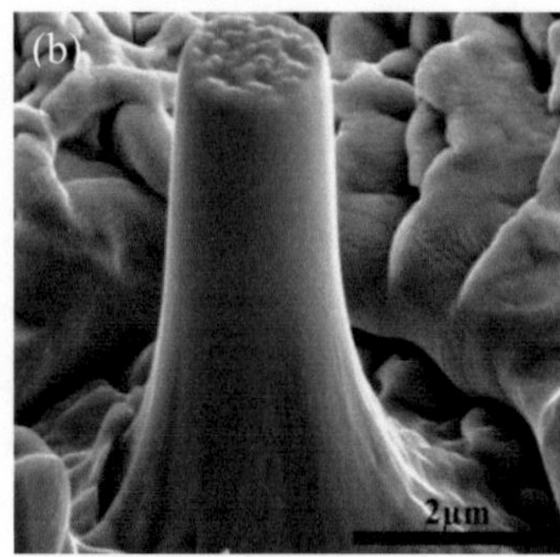

Figure 5.1: Heat treated DC04 steel pillars with diameter of 4 μm (a) and 2μm (b) micro machined by the top down milling method

All micro machined pillars were compressed in the nanoindenter with a constant loading rate. Figure 5.1 shows stress strain data obtained on single crystalline DC04 steel pillars with 2 μm and 4 μm diameter. Even for the same orientation within the same pillars size, considerable scatter in the strength is found. Pillars with diameters of 2 μm show some orientation dependence of their strength where pillars with <111> orientation have the highest strength and show stronger hardening than pillars with <123> and <100> orientations. Among these two orientations, the <100> oriented pillars show the least hardening. For pillars with <100> and <123> orientations, no difference in strength is apparent when comparing pillars with diameters of 2 μm and 4 μm. This is different for the <111> where higher strengths and hardening is apparent for the 2 μm pillars. With decreasing pillar diameter down to 1 μm and 0.5 μm, the stresses increase and the stress-strain data become increasingly disruptive due to strain bursts.

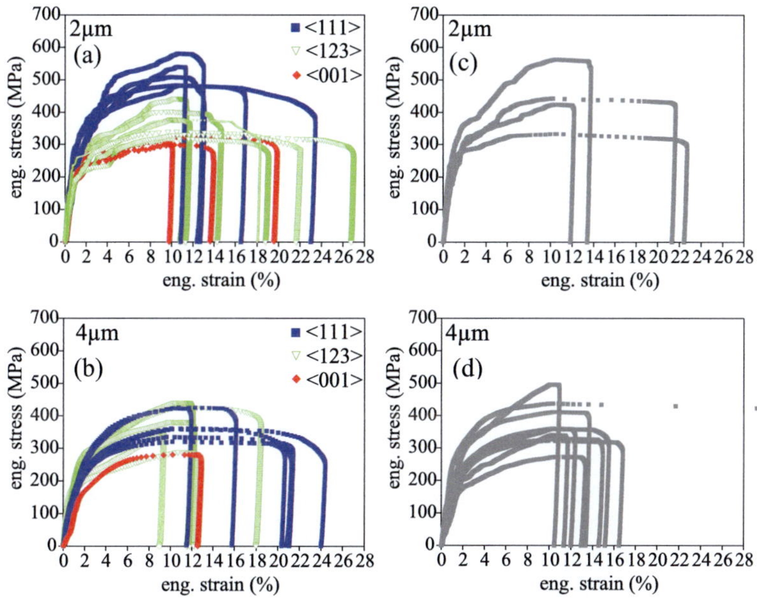

Figure 5.2: Stress-strain data of single crystalline DC04 steel pillars with diameter of 2 μm (a) and 4 μm (b) and bicrystals with diameter of 2 μm (c) and 4 μm (d)

Typical examples of deformed pillars are show in Figure 5.3. A difference in the deformation behavior can be observed for pillars with diameters of 2 μm (Figure 5.3a,c,e) and 4 μm (Figure 5.3b,d,f). Pillars with 2 μm (Figure 5.3a,e) show blurred glide steps after deformation whereas in pillars with 4 μm diameter (Figure 5.3b,f) the glide steps have sharper edges. In general, glide predominantly occurs at the top of the pillar and no clear difference in the deformation morphology can be identified. No localized dislocation glide is found in the pillars. Glide mostly occurs on several glide systems but can also happen on parallel glide planes that seem to belong to the same crystallographic glide system as for example can be seen from Figure 5.4b.

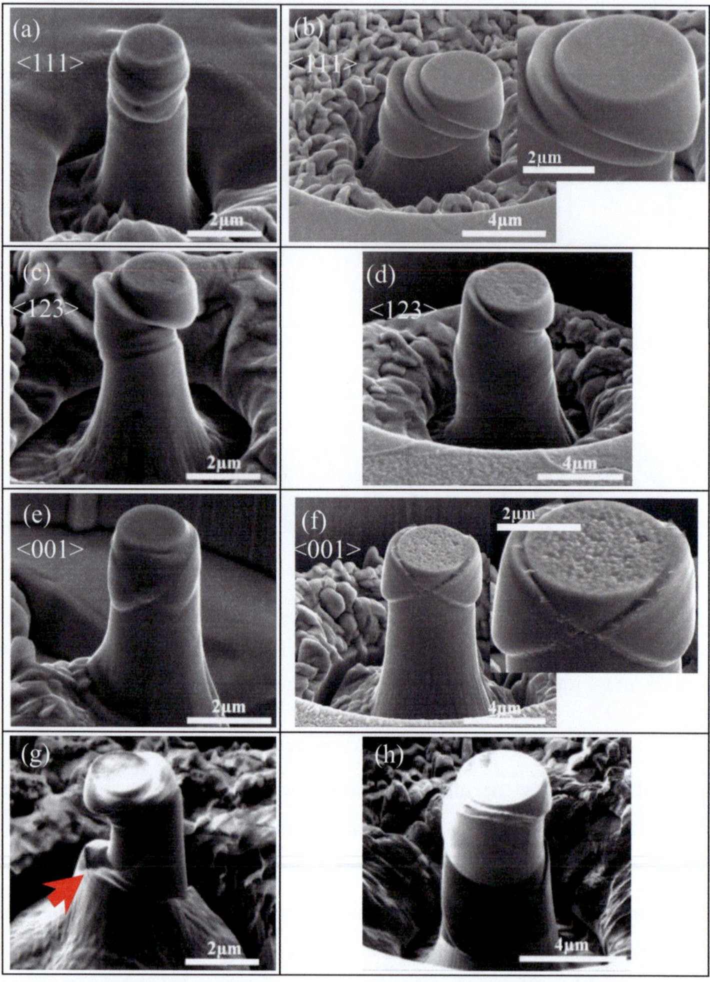

Figure 5.3: SEM images of some of the tested single crystalline DC04 steel pillars with diameters of 2 µm and 4 µm and <111>, <123> and <001> orientations along their axis. FIB images of pillars that contain grain boundaries are shown in (g) and (h). These images indicate that the deformation process is not affected by the grain boundaries. Deformation occurs either inside the grains (h) but can also occur across the grain boundaries (marker in g).

Besides the experiments on single crystalline pillars, bicrystals were tested in order to explore the effect of grain boundaries on the mechanical response of DC04. In general five parameters are required to describe a boundary (the orientation relation between the neighboring grains and the relative orientation of the plane of the boundary) [82], therefore many possible configurations exist and a systematic study would be very time consuming. This was not attempted here, instead boundaries were randomly selected and bicrystals such as the one shown in Figure 5.3g and h were produced and mechanically tested. The morphology of the deformed pillars shows that in all cases slip occurred and that the slip lines did not necessarily end at the grain boundaries. Although the grain boundaries were randomly selected, no strong variation in strength between the different pillars was found (Figure 5.2b and d). In fact, scatter, hardening and strength levels of the bicrystals are not significantly different from what was measured on the single crystalline material in Figure 5.2a and c. In order to further explore the effect of grain boundaries, polycrystalline pillars with a diameter of 22 µm and a height of 60 µm were produced. These experiments are closer to macro scale testing and give information about the influence of grain boundaries on the deformation behavior of DC04 steel. Figure 5.4a shows an FIB image of a pillar after compression. The image shows that deformation in the polycrystal is based on glide inside individual grains but also that glide steps extend across grain boundaries. Figure 5.4 shows stress strain data of three compressed polycrystalline pillars. The curves show a deformation behavior that is relatively similar to what is found for the bulk material. The maximum stresses that were achieved are between 270 MPa and 320 MPa which are in the range of the stresses of the 4 µm pillars shown in Figure 5.2c.

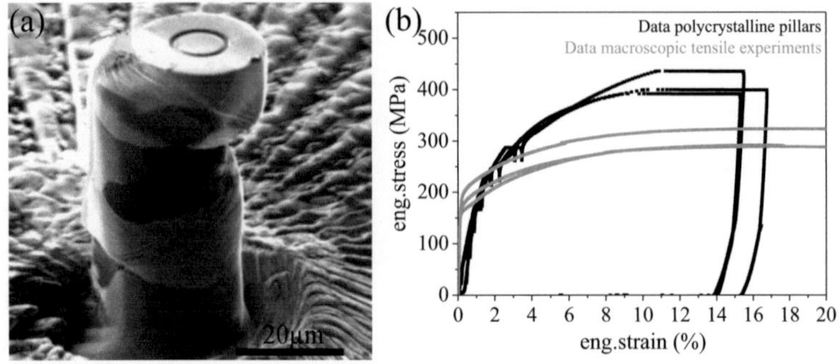

Figure 5.4: FIB image of a polycrystalline DC04 steel pillar after compression (left). Deformation occurs by glide extending across grain boundaries. The right image shows corresponding stress strain data of three compressed polycrystalline DC04 steel pillars and stress strain data obtained from macroscopic tensile experiments on 1.2mm thick sheets of the same material (section 6.1).

From Figure 5.2, it can be seen that the stress-strain response of DC04 steel depends on the pillar size. Besides this data, additional experiments were carried out on pillars with different diameters to investigate the size dependence of DC04. The flow stress data for samples with diameter between 0.5 µm and 22 µm were extracted by drawing a line at 5 % strain parallel to the unloading curve. In Figure 5.5, the flow stress data at 5% plastic strain are plotted for different pillar diameters. The results are compared to the size dependent deformation behavior of α-Fe that was investigated by Daniel Kaufmann [56]. In this study, experiments were performed on single crystal α-Fe pillars with diameter ranging from 0.5 µm to 5 µm (section 2.3.2). The flow stresses of α-Fe follow the described trend with an exponent of -0.81. For diameters below 2µm, the DC04 samples roughly follow this trend and show a very similar scaling. For sample diameters around 2 µm, a relatively large scatter in strength can be seen in the DC04 samples and for pillars that are larger no further decrease in strength is found. This level of strength corresponds to the strength of large polycrystalline samples in compression and in tension (Figure 5.4b).

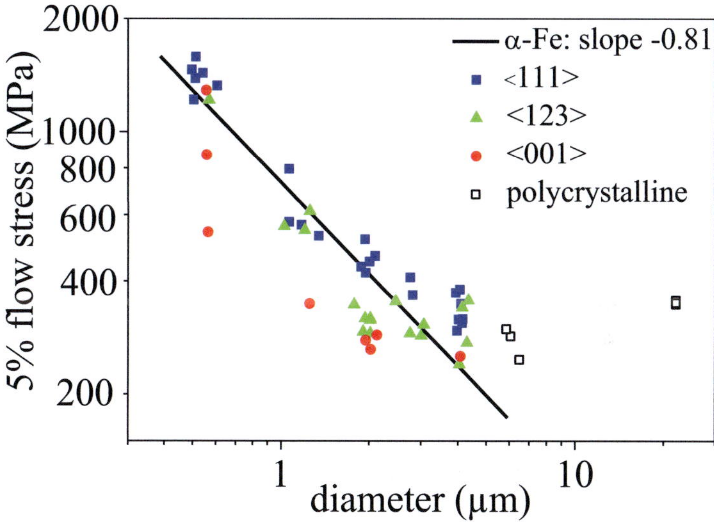

Figure 5.5: 5 % flow stress data plotted to the corresponding pillars diameter. Data for α-Fe were determined by Daniel Kaufmann [56] and are presented in section 2.3.2.

Besides mechanical tests, a detailed microstructural characterization of the heat treated DC04 steel was performed. In cross sections prepared by FIB (Figure 5.6a) small features were observed. The features are located inside the grains and have an estimated average diameter of 0.2 μm and an average spacing of ~2 μm. The distance between these precipitates was estimated based on a FIB slicing of pillars with 4 μm diameter (~12μm height) in 50 nm steps and counting the observed particles in the pillar (Figure 5.6b,c). From this number, the average distance of particles was estimated by dividing the number of particles by the volume of the pillar as determined by the SEM. The edge length of this unit volume was then used as the particle spacing. The EDX spectrum of a precipitate (Figure 5.6d and e) shows peaks that can be referred to Mn and S. Since the precipitate particles have diameters on the order of 200 nm, the EDX spectrum contains signals from the surrounding matrix material. A similar effect is expected for the EBSD patterns (Figure 5.6f-i). In order to correct for it, a pattern of the matrix (Figure 5.6f) and pattern of the precipitate (Figure 5.6g) were taken and the matrix contribution was subtracted so that only the EBSD signal of the precipitate remained (Figure 5.6h). This pattern could then be identified to result from a MnS alabandite phase, a material having a NaCl crystal structure.

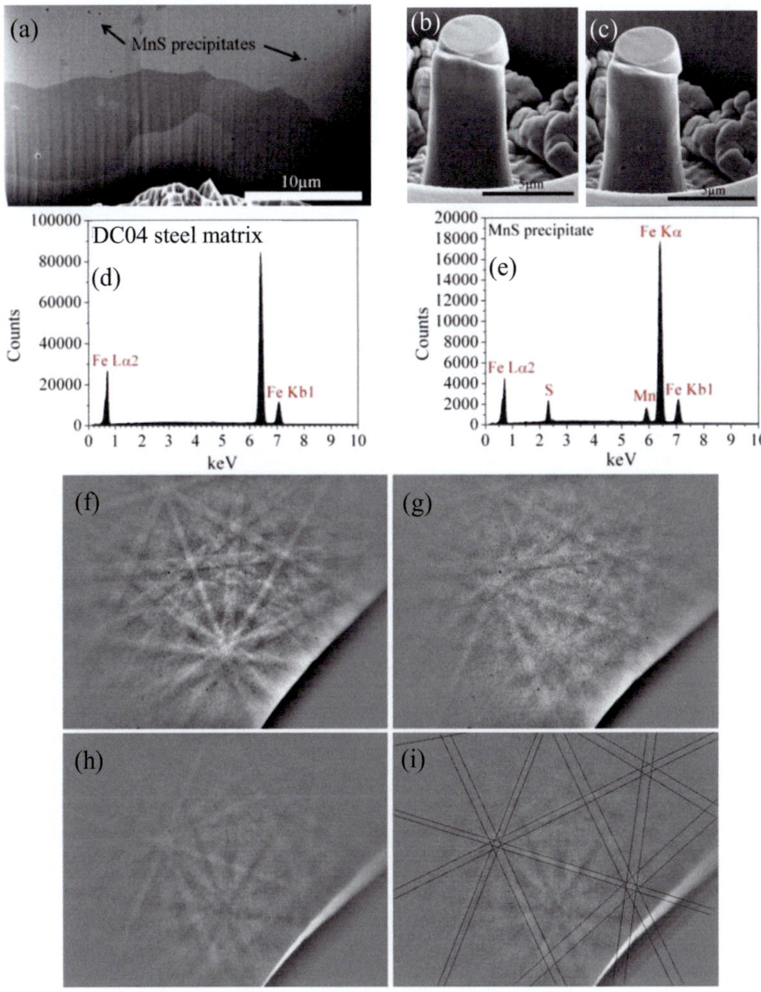

Figure 5.6: SEM image of a cross section of the heat treated DC04 steel (a), FIB cross-sectioning of a pillar (b,c), the composition of the matrix and particles as determined by EDX (d and e) and an EBSD pattern obtained by subtracting patterns of precipitate and matrix (f-i).

5.2　Cold rolled DC04 steel

In the processing route of DC04, a cold rolling process is applied to the sheet metal before the final heat treatment process. It leads to a reduction in sheet thickness from 3.2 mm to 1.2 mm. Microcompression experiments, similar to the heat treated DC04 steel were performed on the cold rolled DC04. Pillars with diameters of 0.5 mm, 1.5 µm and 4 µm were tested. Figure 5.7a and b show stress strain data of the cold rolled DC04 steel pillars with diameters of 1.5 µm and 4 µm. Both pillar sizes show considerable scatter in their strength and the range of the measured stresses is similar for the two different pillar sizes. During deformation, the stress strain data of the 2 µm pillars showed less continuous plastic flow than the larger pillars and pillars made from the heat treated material (Figure 5.2).

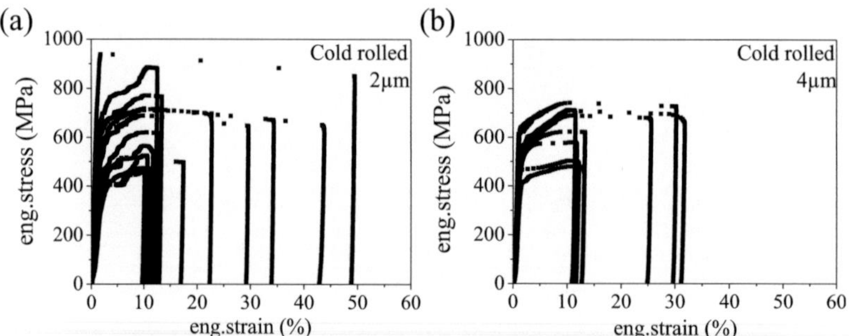

Figure 5.7: Stress strain data of cold rolled DC04 steel pillars with diameters of 2µm (a) and 4µm (b).

Figure 5.8a and b show two representative SEM images of deformed pillars with diameters of 2 µm and 4 µm. In pillars with diameters of 2 µm, deformation is again often located at the top whereas in larger pillars glide processes can also be observed closer to the root. In addition, a FIB image of a compressed pillar in Figure 5.8c shows that the glide processes run through grain boundaries.

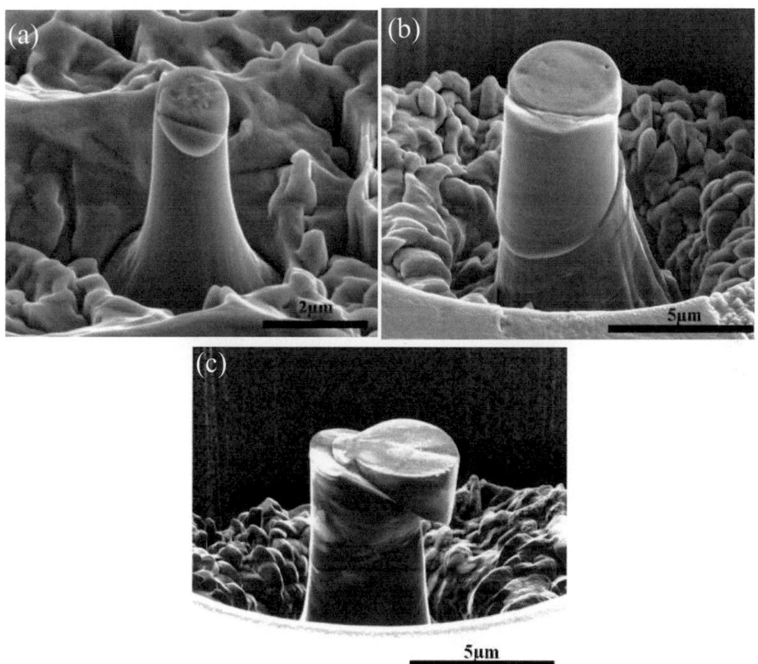

Figure 5.8: Compressed cold rolled DC04 steel pillars with diameters of (a) 2μm and (b) 4μm. (c) FIB image of a compressed 4μm pillar that was micro machined into the cold rolled material.

Figure 5.9a shows an EBSD map of the cold rolled material. It was obtained on a FIB prepared cross section where the cross sectional plane was chosen to be perpendicular to the rolling direction. Due to the cold rolling, the dimensions of the grains are generally larger in the horizontal direction. The variations in color within the individual grains indicate a relatively high dislocation density which was estimated to be in the 10^{14} to 10^{15} $1/m^2$ range.

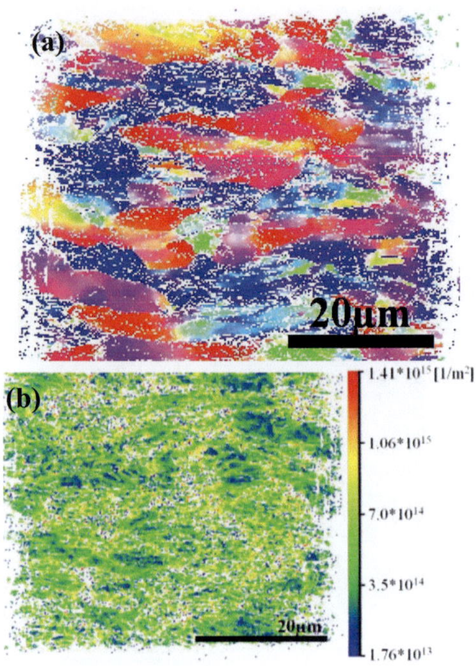

Figure 5.9: (a) EBSD results of a FIB cross section in the cold rolled DC04 steel, Estimated dislocation density from EBSD data are presented in (b). More information on the estimation is given in the discussion.

The 5% flow stresses were extracted from the calculated stress strain data and plotted for the corresponding pillar diameter (Figure 5.10). The data show that the cold rolling leads to a decrease of the flow stresses for pillar sizes smaller than 1-2 µm compared to α-Fe and heat treated DC04. Pillars larger than 2 µm are stronger the samples that were micro machined into the heat treated DC04 steel or α-Fe. Overall, a size effect is not discernible in the cold rolled material.

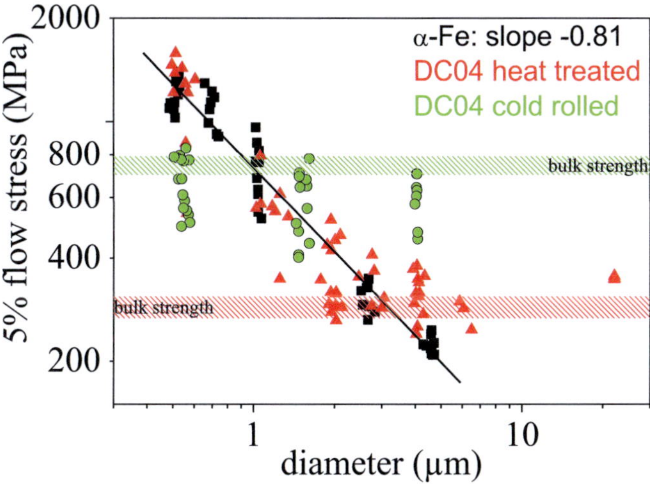

Figure 5.10: 5% flow stress plotted to the corresponding pillars diameter for the heat treated and cold rolled DC04 steel. Data for a-Fe were determined by Daniel Kaufmann [56] and are presented in section 2.3.2.

5.3   Discussion

The mechanical behavior of α-Fe (section 2.3.2) and DC04 shows similarities in many aspects. The shape of the stress strain curves and their size dependence are very similar for pillars with diameters below 2 µm. Also, the deformation behavior follows the trend known from many metals, namely that with decreasing pillar size the strength increases and that the stress strain response transitions from continuous flow to individual strain bursts.

5.3.1 Heat treated DC04 steel

Small DC04 pillars with diameters below 2 μm, show stress-strain responses and strengths that are similar to that of α-Fe pillars (Figure 2.14b). Deviations from the behavior of <211> α-Fe occur for larger pillars. From Figure 5.4 and Figure 5.5 it can be seen that the strength of the heat treated DC04 steel pillars with diameters larger than 4 μm show stresses that are close to the bulk strength of this material which is around 290-320 MPa. In contrast to α-Fe, this bulk strength is rather high. Pure single crystals of α-Fe can have strengths as low as 10 MPa [65, 67]. Several reasons are possible for the enhanced strength of DC04 as compared to α-Fe. For example the dissolved carbon can have a strong influence on the mechanical properties of the material due to interstitially dissolved carbon atoms and the interaction of the resulting stress fields with dislocations [36, 67]. For α-Fe that is alloyed with carbon, an increase of the strength can be observed at room temperature which is proportional to the carbon concentration (section 2.3.1). A carbon content of ~0.001 wt% increases the strength about ~100 MPa [36]. Therefore, it is most likely that the carbon content in DC04 steel increases the strength by 300 MPa. This hardening is expected to be size independent and should lead to the same amount of strengthening for samples of different size. For pillars sizes smaller than 2 μm a strengthening similar to α-Fe can be found due to size effects that control the strength of the pillars in this size regime. This size effect strengthening is stronger than the hardening due to interstitially dissolved carbon and is therefore the dominant mechanism for pillar sizes smaller than 2 μm.

Moreover, the microstructural characterization of DC04 steel showed that the material contains small precipitates that were analyzed to be MnS (Figure 5.6). These presumably incoherent particles can serve as obstacles that hinder dislocation motion. The Orowan law $\Delta\sigma = Gb/L$ can be used to calculate the increase in flow stress caused by hard obstacles. The measured average particle distance within the grains is $L = 2$ µm, the shear modulus of α-Fe is $G=82.4$ GPa and the lattice constant $a \approx 2.9$ Å. According to the Orowan law, these values result in an increase in strength of ~10 MPa. With this simple estimation only a small fraction of the strength can be explained.

Another possibility is that size effects may be responsible for the observed increase in strength. From thin films it is known that the strength of the film increases with decreasing film thickness [57, 95, 96]. A similar effect can be found in micro- or nanocrystalline materials. According to the Hall Petch law, the strength of the material is inversely proportional to the square root of the grain size [37, 57]. These two very different systems show that strengthening mechanisms exist that scale with sample or feature size. In particular when dimensions approach the sub micrometer scale, strong scaling effects are often found. This may be due to the fact that instead of dislocation networks only individual dislocations carry plasticity and that the availability of mobile dislocations decreases with decreasing sample or feature size. Since these concepts seem to be valid for varying geometries ranging from pillars over films to nanocrystalline materials it is not implausible that similar concepts may also be applicable to particle strengthened materials when the spacing between the particles reaches into the micrometer scale. In the DC04 steel investigated here, the average particle spacing is roughly 2 µm. Therefore, pillars with diameters larger than 2 µm should contain more than one particle. If the particles are located on the glide planes of the moving dislocations, they serve as obstacles that hinder dislocation motion and cause hardening. In this case, a strengthening of these pillars would be expected. This strengthening would not depend on pillars size as long as the pillar contains one or more particles that are located on the glide planes of the moving dislocations. In this case, the shortest dimension (including particle spacing or spacing between particles and pillar sidewall) would control the strength which may lead to the observed strengths of around 320 MPa. In pillars where the diameter is smaller than the particle spacing, i.e. smaller than 2 µm, the strength may be controlled by the pillar dimensions and a size scaling of these "particle free" pillars similar to the one of α-Fe is expected and observed (see Figure 5.5). It seems plausible that in the transition region between size controlled strengthening and particle spacing controlled strengthening, a large scatter in strength would occur due to variations in particle number and position inside the individual pillars. This concept is supported by micro compression data from ODS strengthened nickel [97]. There the strength of the ODS nickel was around 500 MPa and did not strongly scale with pillar size, consistent with a particle spacing of ~100 nm as observed by TEM.

For pillar diameters of 2 µm, a large scatter in the strengths can be seen in Figure 5.5. A closer inspection shows that the <111> oriented pillars have a higher strength than the other pillars. The strengths fall onto the trend line that describes the size scaling of the <211> oriented α-Fe. Pillars with diameter of 2 µm and <123> or <001> orientations have slightly lower strengths that are roughly in the range of the bulk strength of DC04. With decreasing pillar diameter, the strengths of the <123> oriented pillars also increase and reach values similar to the <211> α-Fe and <111> DC04 samples. The <100> oriented DC04 samples show a behavior that is markedly different from the rest of the data. The onset to size dependent behavior occurs at smaller pillar size and their strengths stay below the strengths of the other orientations. Overall, the size scaling of the strength of DC04 seems to depend on the orientation (Figure 5.11a). Not only the onset of the size dependent strengthening occurs at different sizes for different orientations, but also the scaling of the tested pillars depends on their orientation.

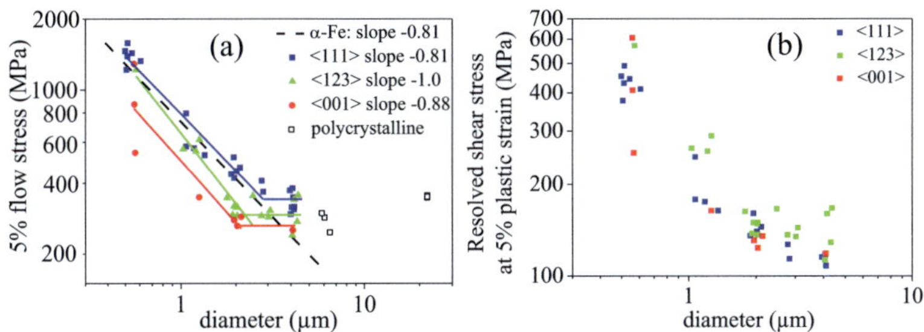

Figure 5.11: (a) Estimated size scaling exponents of the strength for DC04 steel pillars with <001>, <111> and <123> orientations and possible trendlines of the stresses that indicate the different onsets of the size dependent behavior. (b) Resolved shear stresses in the single crystalline pillars calculated at 5 % plastic strain with Schmid factors for slip on {211} (Table 1).

Glide planes	Schmid factors m	Loading directions		
		<001>	<111>	<123>
{110}	highest m	0.41	0.27	0.47
	second highest m	0	0	0.35
{211}	highest m	0.47	0.31	0.47
	second highest m	0.24	0.16	0.34

Table 5.1: Schmid factors m for different orientations assuming glide on {110} or {211} planes with a burgers vector of <111>

Surprising are the high strengths that are found for the <123> oriented pillars. According to the Schmid factors in Table 5.1, the strength values are expected to lie not far from the <100> data. The high strengths found in the experiments could be due to another mechanism. <100> and <111> orientations have high symmetry with eight or six equivalent slip systems that all have the highest Schmid factors. In <123> pillars, only one glide system with the highest Schmid factor is available. If the mechanical strengthening of small samples is due to the limited availability of dislocations as often discussed [46, 57, 98], then crystal symmetry may play a role. In the high symmetry pillars, more possibilities exist for nucleating dislocations than in the single slip configuration [75]. Such an argument could explain the rather high strengths that were found in the small single slip <123> pillars.

Characteristic deformation morphologies of pillars are depicted in Figure 5.3. There is a change in morphology between 2 μm and 4 μm pillars. Independent of their orientations, pillars with diameters of 2 μm predominantly deform at the top whereas 4 μm pillars also tend to deform closer to the root as it was the case in Figure 5.3b,d,f. Such an effect may be related to stress concentrations at the interface between tip and pillars. At a given stress on the pillar, the sample edges or a surface irregularity would lead to stress concentrations. The field of this stress has characteristic dimensions that are not necessarily related to the sample dimensions. If the samples are large compared to this stress field, mostly uniform straining will happen. For small samples, the stress concentration may dominate and lead to a localization of the plastic deformation as observed on the 2 μm pillars. Comparing deformation morphologies between different orientations and sizes suggests that the large <111> pillars (Figure 5.3b) deform differently than the others. On these pillars, very distinct slip processes on parallel glide planes were found and from their spacing it seems that only little or no interaction between different glide systems occurs. These observations indicate that the 4 μm pillars with <111> orientation are less resistant to glide than the other samples. This assumption is fortified by the stress strain data shown in Figure 5.2b where data was recorded at constant rate and therefore the spacing between data points inversely correlates to the rate of deformation. The highest deformation rates and also the largest plastic strains are observed for the <111> pillars. In contrast to this, the stress strain data of pillars with <123> orientation show that the samples are continuously hardening during compression. This may also be inferred from the deformation morphology of the <123> pillar in Figure 5.3c where it can be seen that the glide steps are not as distinct and that often more than one glide system was active. The activation of the second glide system in <123> pillars can be explained by the relatively high Schmid factor of the second glide system. When comparing the shapes of the 2 μm with the 4 μm pillars it can be seen that there is a general trend that the slip steps are more blurred in the smaller samples. This suggests that in the larger samples, glide is more localized and happens on individual planes and that in the small samples deformation is less localized and happens on neighboring planes.

The deformation morphologies of polycrystalline pillars (Figure 5.3 g,h) are not very different from the ones of single crystalline pillars. In the polycrystals, slip traces sometimes run across grain boundaries as can be seen in Figure 5.4. This is surprising since the orientations of the grains vary strongly. The bcc crystal has 24 or 48 glide system, depending if the {123} glide planes are included. These large numbers of glide systems may facilitate the continuation of slip in contiguous regions with different crystal symmetry. The fact that grain boundaries are no or only weak obstacles to dislocation motion is apparent in the stress-strain responses in Figure 5.2. When comparing Figure 5.2a,b with Figure 5.2c,d it can be seen that single and polycrystalline samples mechanically do not behave significantly different. When considering samples with many grains, the strength and the stress strain response of polycrystalline pillars with diameter of 22 μm shows some similarity to that of macroscopic DC04 samples (Figure 3.2). The elastic slope measured on the polycrystalline pillars is too low. This is a common phenomenon observed in microcompression experiments and may be attributed to the onset of localized plasticity, stress inhomogeneities directly underneath the flat punch, misalignment between flat punch and pillar or the elastic deformation of the material underneath the pillar. Therefore, the elastic modulus should be determined from unloading data. The fact that the measured flow stresses in these macroscopic samples do not deviate from the flow stresses measured in tensile experiments and in the 4 µm single crystals suggests that the presence of grains in this size regime does not lead to significant strengthening in DC04.

5.3.2 Cold rolled DC04 steel

In the steel process chain, cold rolled sheets are heat treated resulting in the material that has been investigated so far. In the following, the cold rolled material before the heating step will be addressed. During cold rolling the thickness of this material was reduced by a rolling reduction of 63%. Figure 5.9a shows EBSD data of a FIB cross section that was micro machined into this material. The shades in the colors within the grains indicate a high density of dislocations. The density of geometrically necessary dislocation (GNDs) ρ was roughly estimated by using the EBSD data shown in Figure 5.9a and applying these data to equation (2.5). It has to be emphasized that this treatment only gives estimates for the dislocation density and does not account for the different slip systems. Such a treatment requires a significantly more complicated analysis [17, 99]. Despite the simplicity of this estimation it was shown in the literature that both, this estimation and the more elaborate concept yielded very similar results for a dual phase steel [17].

In Figure 5.7 the stress-strain data of pillars with a diameter of 4 µm are depicted. Compared to α-Fe and the heat treated DC04 steel, the maximum stresses are roughly twice as high. It is straightforward to argue that the higher dislocation density leads to dislocation-dislocation interactions that are stronger in the cold rolled material. The increase of the bulk strength due to cold work hardening is commonly assumed to be proportional to the square root of the dislocation density $\Delta\tau = \alpha Gb \cdot r^{1/2}$ according to the Taylor equation with $\alpha = 0.4$ for bcc metals. Following this, a dislocation density of $9*10^{14}/m^2$ which is the average value of estimated dislocation density would increase the strength by 245 MPa. This roughly agrees with the measured difference of ~280 MPa between the strength of the heat treated DC04 steel pillars and the cold rolled pillars for a pillar diameter of 4 µm (Figure 5.10). It should be noted that the data is only based on a rough estimate of the dislocation density and additional statistically stored dislocations that cannot be detected using EBSD with a step size of 200 nm may also contribute to the strengthening. Also, the dislocation density map in Figure 5.9b shows regions of increased density which may be related to cell structures that form during rolling. In the literature, the development of the microstructure and the cell sizes was investigated in an interstitial free steel after different rolling reductions by Godfrey et al [38]. After a rolling reduction of 70 % the cell sizes were smaller than 500 nm. The features in Figure 5.9b are not strictly periodic but seem to have dimensions in the low micrometer range. The dislocation networks that exist in the cold rolled material represent obstacles for moving dislocations which lead to higher strengths as compared to the low defect heat treated material.

Smaller pillars with diameters of 2 µm (Figure 5.7 a) show stress strain responses that are similar to that of the 4 µm pillars (Figure 5.7 b) with levels of flow stresses comparable to α-Fe at 2 µm. In Figure 5.10 the flow stress data are plotted for different pillar sizes. The flow stresses in the cold rolled material changes only little with decreasing pillar diameter. As a consequence, pillars with diameter of 0.5 µm have flow stresses that are significantly lower than that of the heat treated DC04 steel and α-Fe. This finding will be discussed in the light of a number of theories that try to explain the size dependent strength of metal single crystals. Dislocation starvation is one possible explanation [46, 47, 57, 98]. With decreasing sample size, the number of dislocation sources in the material decreases. In addition, the surface-to-volume ratio of the pillars increases which leads to higher probabilities to annihilate moving dislocation at the nearby surface due to image forces. In this situation, deformation is limited and controlled by the availability of dislocations. In the absence of dislocations, high stresses may be required for their nucleation leading to a strengthening. In the cold rolled DC04 steel, a high dislocation and defect density is present which may alleviate dislocation nucleation. At these defects, additional mechanisms for dislocation nucleation can take place that require lower stresses. This finding is in agreement with the measurements bei Bei et al. on pre-strained Mo-based pillars who found a decreasing strength of 500 nm pillars with increasing pre-strain [72].

In Figure 5.8 the deformation morphologies of pillars that were micro machined into the cold rolled DC04 steel are depicted. The contrast in the FIB image (Figure 5.8c) shows that the pillars contain regions with different orientations and therefore several grain boundaries. During compression, the active glide processes also seem to cross the existing grain boundaries. This observation fortifies the assumption that grains boundaries do not significantly affect the deformation behavior in DC04 steel both in the heat treated and in the cold rolled material. In pillars with 4 µm diameter, small slip steps can be observed at the surface of the pillar. This indicates that during deformation of the cold rolled material glide processes are not localized onto single glide systems but that glide processes can be activated on several glide planes.

In order to summarize the results of this chapter, Figure 5.12 illustrates the individual factors and their effect on the size scaling of strength. It is demonstrated that the size scaling is affected by many parameters and that it is not justified to describe it by a single exponential relationship from a mechanistic point of view.

In the heat treated DC04 steel, pillars with diameters larger than 2 µm show strengths that are higher than what was found for α-Fe. These strengths are in agreement to the bulk strength as measured by tensile experiments on macroscopic samples. The increase of the strength compared to pure Fe may is most likely caused by interstitially dissolved carbon and/or particle strengthening of small MnS precipitates that were found in the matrix, respectively. For pillars with diameters smaller than 2 µm, a size dependent deformation behavior similar to α-Fe was found. Additionally, it seems that the strengths and the scaling exponents dependent on the orientation of the pillars. Pillars with <111> and <123> orientations show a similar scaling as α-Fe whereas pillars with <001> orientations have lower strengths and a scaling with a smaller exponent. The different strengths were explained by different resolved shear stresses assuming that glide can occur on {011} and {112} planes. The different scaling may be explained by the fact that more nucleation sites exist in high symmetry orientations than in pillars oriented for single slip. The onset of the size dependent behavior also seems to be orientation dependent and occurs at the smallest sizes for <100> oriented pillars. The effect of residual dislocations was investigated by experiments with cold rolled D04 steel pillars that contain a high density of dislocation. For pillars with a diameter of 4 µm the strengths are higher than what was determined for the heat treated material. This increase is explained by work hardening and possibly the formation of a cell structure in this highly deformed material. In the cold rolled steel, almost no size dependence of the strength was found. It is argued that this observation is in agreement with the dislocation starvation theory since dislocation nucleation may not be a limiting factor for small pillars that already contain a high number of dislocations.

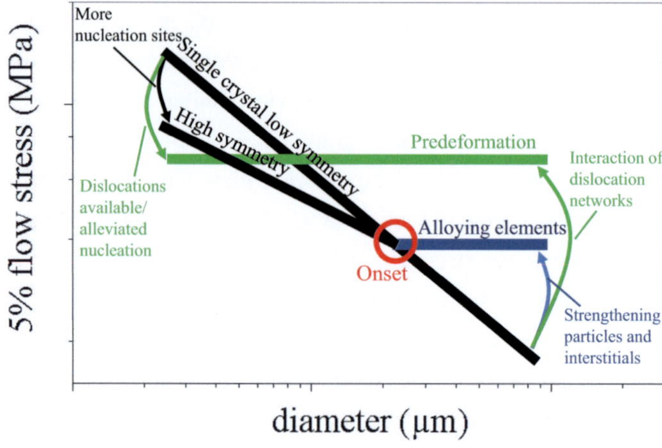

Figure 5.12: Summary of effects that may affect the size scaling of the mechanical strength of α-Fe and low alloyed steel.

6. Deformation Behavior and Microstructure Evolution During Tensile Experiments

In the Graduiertenschule 1483 deep drawing processes are simulated with the use of micro mechanical models (Project A2). These micro mechanical models are developed and verified by the use of experimental data that were received from tensile experiments and corresponding microstructural data before and during deformation. This chapter contains results of macroscopic tensile experiments that were performed on heat treated DC04 steel. Besides conventional tensile tests on DC04 samples, special experiments were carried out to investigate the evolution of the microstructure during straining. The mechanical experiments were interrupted at predefined amounts of total strain. At each interrupt the sample was unloaded and characterized by EBSD before further testing was performed.

6.1 Anisotropic deformation of DC04 sheet metal

In order to investigate the macroscopic deformation behavior of heat treated DC04 steel, tensile experiments were performed with samples that were aligned parallel, 45° and 90° to the rolling direction. These experiments were the first experiments that were performed with this Ph.D. work. At that time the material directly coming from the production process was not available and therefore the experiments were performed on a different batch of DC04 steel as compared to the rest of the experiments presented in this thesis. Slight differences in composition may be possible and therefore deviations in the deformation behavior may occur. Results of the tensile experiments are shown in Figure 6.1.

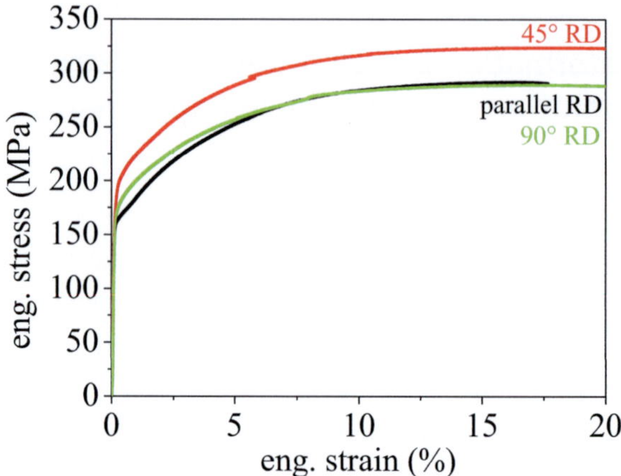

Figure 6.1: Stress strain data of DC04 steel samples that were aligned parallel, 45° and 90° to the rolling direction

The samples were not strained to the fracture strain since the experiment was used to characterize the microstructure and therefore the straining was stopped at 20%. The data clearly show, that the material exhibits higher yield stresses and tensile strengths when the sample is loaded at an angle of 45° relative to the rolling direction and that hardening is not very different for the different directions.

6.2   Evolution of the microstructure during tensile straining

Similar tensile experiments were performed on the standard heat treated DC04 steel that is used throughout this thesis. These experiments were performed on two samples that were aligned parallel to the rolling direction. The first sample was strained until fracture and the corresponding stress-strain data were monitored. The second sample was unloaded after 1 %, 5 %, 12 %, 19 % and 28 % plastic strain and the microstructure was recorded by EBSD. Before testing, platinum markers were deposited on the surface of this sample using FIB [100], and used for measuring the residual plastic strain after each unloading step by SEM. In addition small dust particles served as markers to measure the residual plastic strain. Figure 6.2a shows an SEM image of the sample before straining with Pt markers and dirt particles (e.g. in the center). Figure 6.2a and b show the sample surface before and after 28 % of plastic strain. The green lines indicate how the plastic strain was determined.

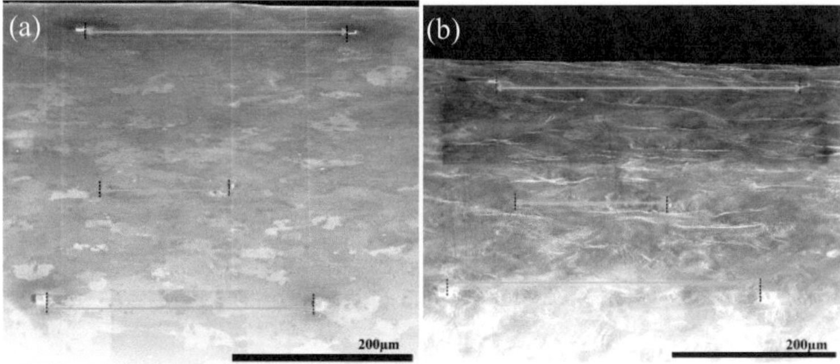

Figure 6.2: SEM images taken at 70° tilt (necessary for EBSD). Image (a) is the inital state. At the ends of the green lines are markers that are either Pt depositions or dust particles. Image (b) is the sample after 28 % plastic strain.

Figure 6.3 shows the stress-displacement response of the heat treated DC04 steel with a yield strength of around 180 MPa and a tensile strength of around 325 MPa. The second sample was strained to increasing loads in five steps. The sample was unloaded after 140 N, which is close to the yield strength, after 0.6 mm, 1.4 mm, 1.1 mm and 1.8 mm. Table 6.1, contains the plastic strains as determined by SEM on three positions (Figure 6.2).

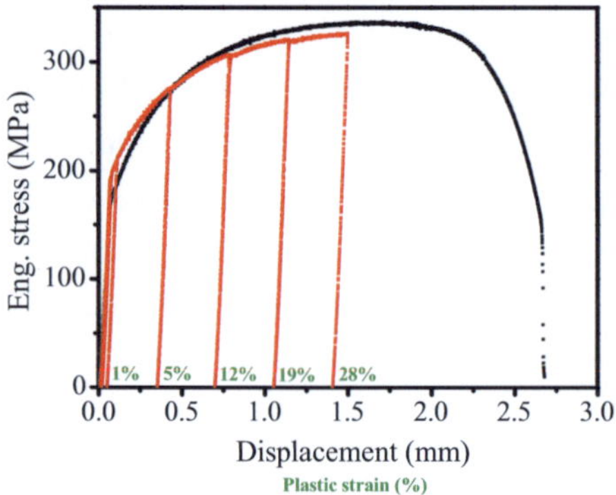

Figure 6.3: Stress-strain data of tensile specimen loaded parallel the rolling direction

Marker 1	1.3%	5.1%	11.4%	18.2%	26.9%
Dirt particles	0.9%	5.2%	12.3%	19.3%	28.1%
Marker 2	1.4%	5.5%	11.4%	18.9%	27.8%
Mean	1.2%	5.3%	11.7%	18.8%	27.6%

Table 6.1: Results of the strain measurements by analyzing the marker distance on the sample surface after five unloading steps

Figure 6.4 shows SEM images of the initial state and after 1 %, 5 %, 12 %, 19 % and 28 % plastic strain. All images were taken at the same magnification. The dark grey areas that can be observed on the sample surface result from the EBSD measurements where the electron beam left carbon contaminated regions behind. During the tensile experiment the sample is strained in the horizontal-direction and therefore the EBSD area becomes longer and smaller in height. In images Figure 6.4c-f only the inner parts of the EBSD region remains visible. The images clearly show how the sample increasingly became rough and that more and more thickness variations and glide steps appeared with increasing strain.

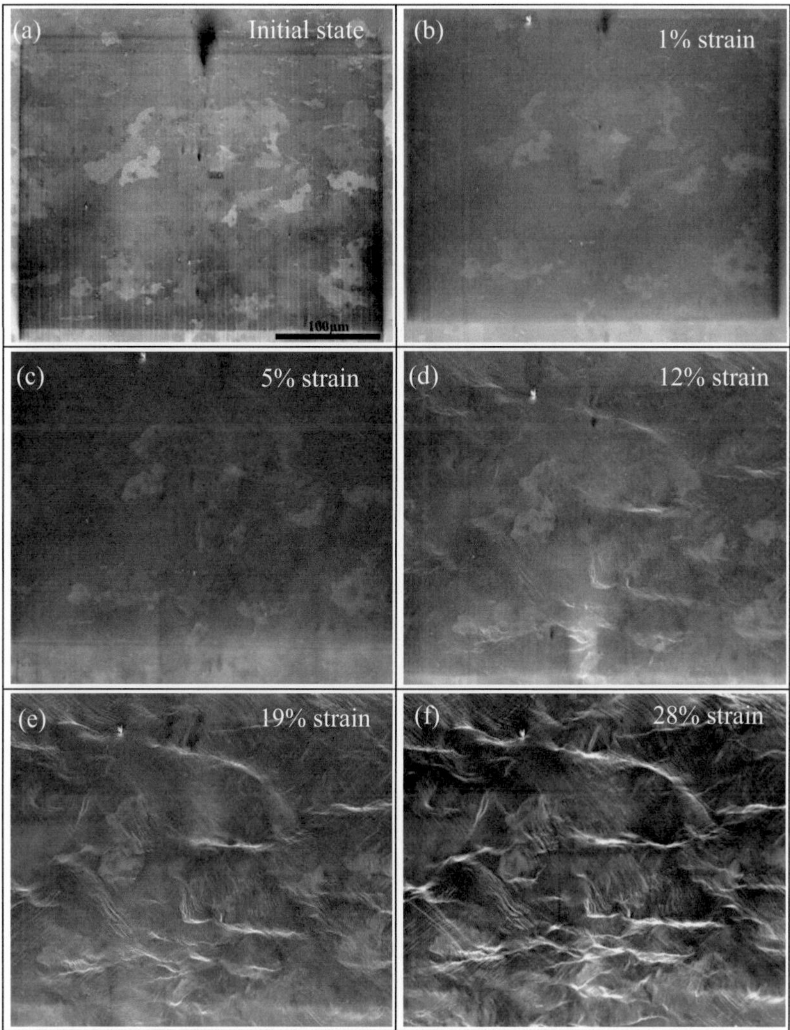

Figure 6.4: SEM images of the sample at the initial state and after 1 %, 5 %, 12 %, 18 % and 28 % plastic strain. All images were taken at the same magnification.

To investigate the local changes of the crystal orientation during the tensile experiments, EBSD measurements were performed after each unloading. Figure 6.5 shows the EBSD maps of the unstrained sample and after each of the five unloading steps. The EBSD maps were plotted in two directions of the sample coordinate system, along the sheet normal and along to the tensile direction. After a plastic strain of 1 %, the EBSD maps show similar crystal orientations and grain shapes than the starting material. After 5 % plastic strain the crystal orientations start to change within the grains due to the deformation of the material but the shape of the grains does not change considerably. In the corresponding SEM image in Figure 6.5 first glide steps became visible after 5 % of plastic strain. After further deformation, the crystal orientations continue to change within the grains and the grain shapes start to deform and elongate along the tensile direction. After 20 % of plastic strain, the EBSD maps show larger regions in which the crystal orientations could not be determined anymore (white pixels) due to the roughness of the sample surface and the high dislocation densities that distort the Kikuchi pattern and lead to poor indexing during EBSD mapping.

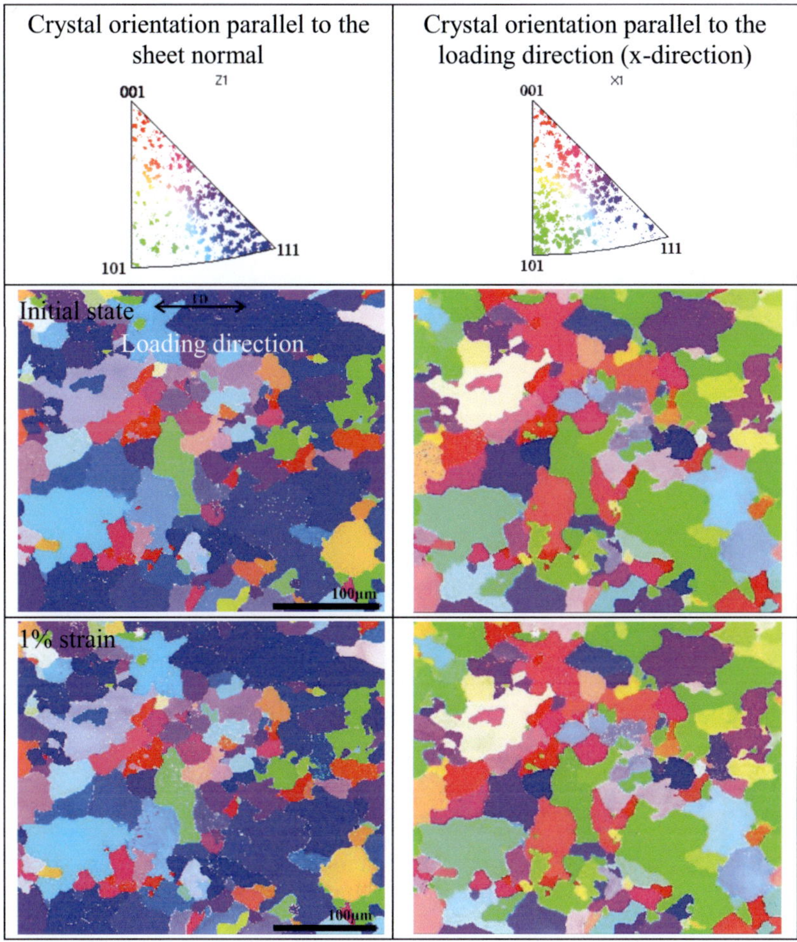
Crystal orientation parallel to the
sheet normal
001
Z1
101
111
Crystal orientation parallel to the
loading direction (x-direction)
001
X1
101
111
Initial state
LD
Loading direction
100μm
1% strain
100μm

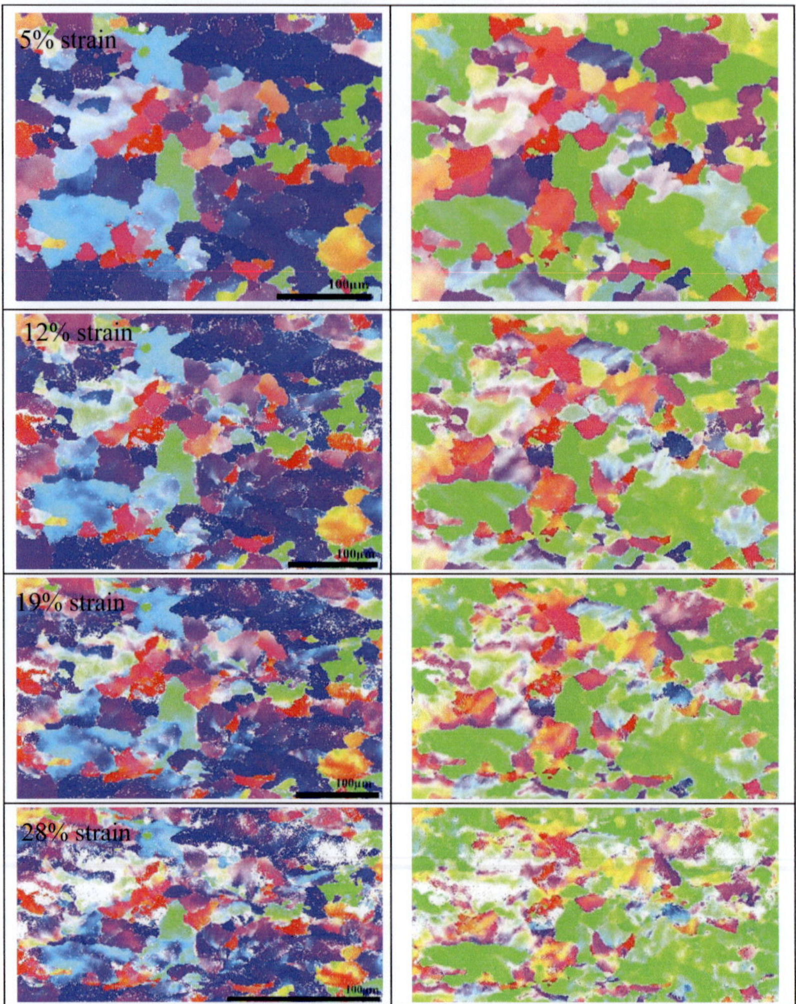

Figure 6.5: Inverse pole figure map of the sample before and after 1 %, 5 %, 12 %, 18 % and 28 % residual plastic strain. The color coded inverse pole figure of the initial state is plotted. It may be used as a reference to correlate colors with orientations.

The EBSD results shown in Figure 6.5 were also analyzed by plotting the {111}, {110} and {001} pole figures (Figure 6.6). The pole figures were calculated based on the EBSD data depicted in Figure 6.5. In the initial state, the non deformed material consists of discrete orientations (individual grains) but after 28 % strain large misorientations exist and the poles are not distinct so that it becomes difficult to analyze the individual texture components. Here a different approach is taken. In order to additionally obtain spatial information on individual texture components, maps of the microstructure (Figure 6.5) were used to track the evolution of the grains.

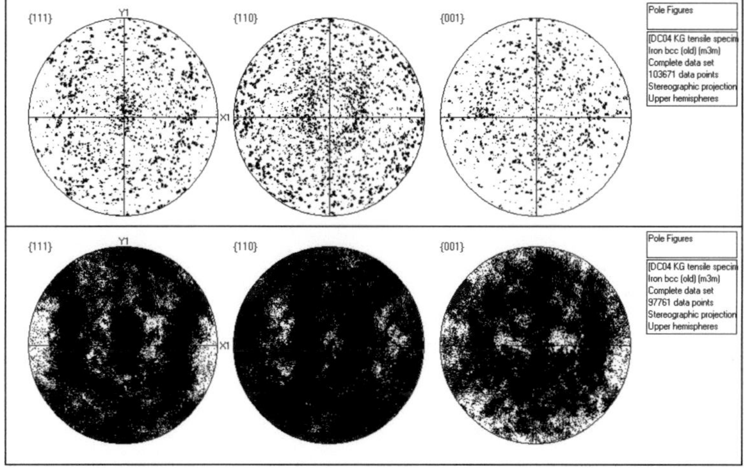

Figure 6.6: Calculated pole figure for the {111}, {110} and {001} planes of the initial state and after 28 % residual strain

The pole figures of the heat treated DC04 steels in section 4.1.1 and section 4.1.2 show that the texture has α- and γ- fiber components, which implies that the <111> orientations are preferentially aligned parallel to the sheet normal and the <110> orientations are aligned parallel to the rolling direction. Furthermore, the pole figures of the cold rolled DC04 steel have dominant <100> orientations that are aligned parallel to the sheet normal. For this reason the evolution of these main texture components (<111>‖ND, <110>‖RD and <001>‖ND) during the tensile experiments are investigated in detail. Figure 6.7 shows the <111> and <001> out of plane orientations of the initial state and after 12 % and 28 % stain and the <110> orientations that are aligned along the tensile direction.

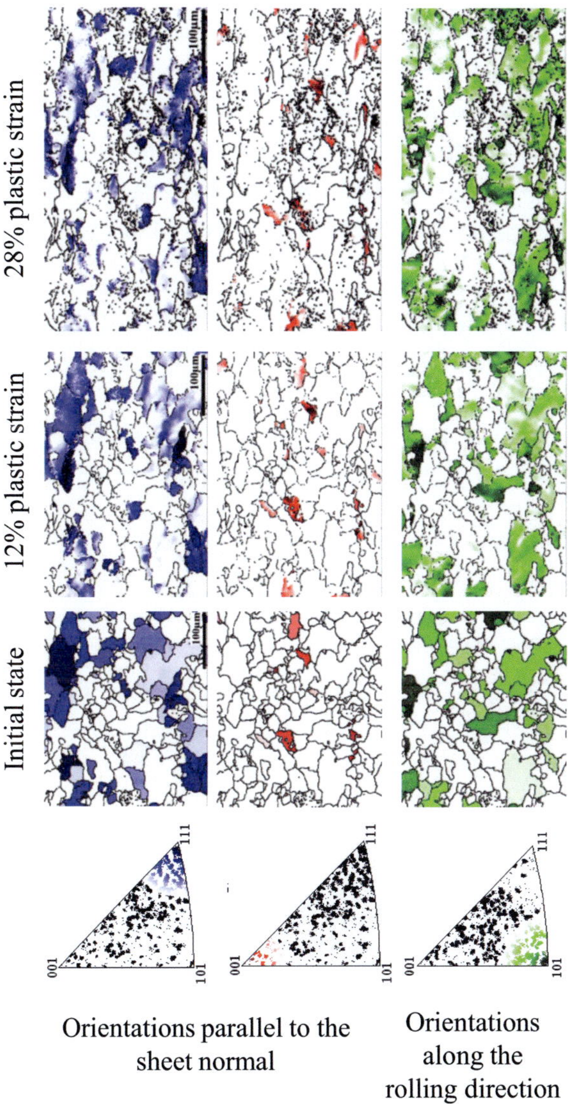

Figure 6.7: The <111> and <001> orientations are plotted with respect to the sheet normal and the <110> orientations parallel to the tensile direction. All orientations lie within 15° misorientation to the corresponding direction

The area fractions of the three components were analyzed after each unloading step and the results are plotted in Figure 6.8a and b. During straining, the area fractions of the <111> out of plane orientations decrease by ~12 % whereas the fraction of <001> orientations increase by ~2 %. The strongest change happens in the area fractions of the <110> orientations along the tensile direction (Figure 6.8b) which increase about roughly 20 %.

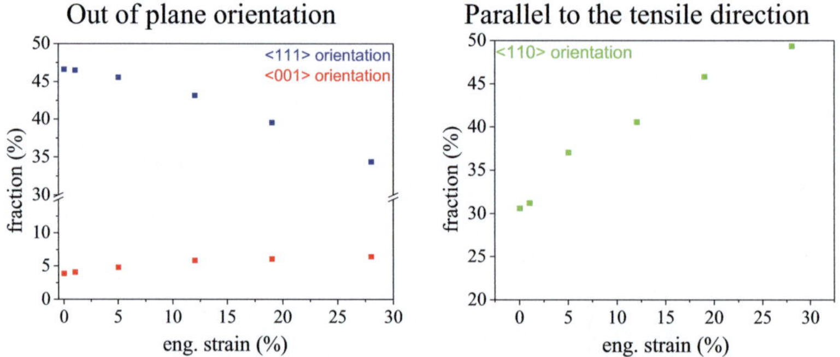

Figure 6.8: (a) shows the development of the area fraction of <111> and <001> orientations parallel to the sheet normal. (b) is the area fraction of <110> orientations aligned along the tensile direction.

A detailed analysis of the orientations on the α- and the γ-fibers are depicted in Figure 6.9. The results show that the {111} <110> and {112} <110> texture components increase during tensile experiments whereas the {111} <123> and {111} <112> components decrease. In other words, the overall density of the α-fiber increases and the orientation density in the γ-fiber decreases.

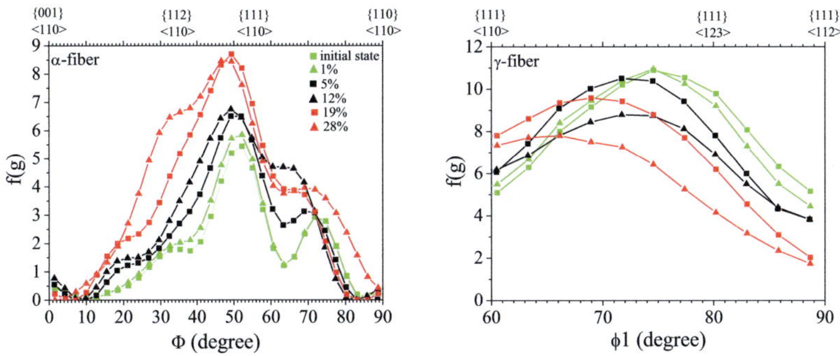

Figure 6.9 ODF plots that show the α-fiber and γ-fiber of the initial heat treated DC04 steel and after 1%, 5%, 12%, 19% and 28% plastic strain

Besides investigations on the evolution of orientation during straining additional measurements on the evolution of misorientations within texture components were performed. These analyses were done with grains that have their <111>, <110> and <001> orientations aligned along the sheet normal and with grains having these crystallographic orientations along the tensile direction. Grains with these orientations were determined in the initial material within an angle of 15° and the development of their average misorientation was tracked after each unloading step. Figure 6.10 shows that the overall average misorientation increases after each unloading step. For the orientations parallel to the sheet normal no differences between the <111>, <110> and <001> orientations can be observed. The average misorientation at the end of the test is about 3° for all orientations. Grains with <111> orientations along the tensile axis show higher misorientation angles than grains that have <001> and <110> orientations.

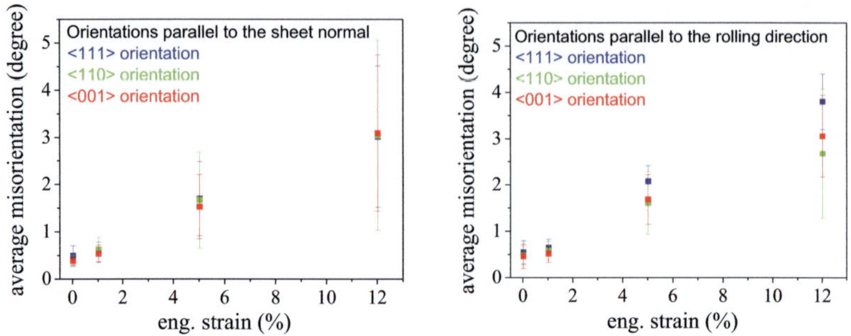

Figure 6.10: Changes of the average misorientation tracked in grains with <111>, <110> and <001> orientations that are aligned parallel to the sheet normal and parallel to the tensile direction.

The misorientation between the grid points of the EBSD area can further be used to calculate the density of geometrically necessary dislocation (equation (2.5)). Figure 6.11 shows the density of the calculated GNDs for the untested heat treated DC04 steel and its development over the five unloading steps. Before deformation took place the material had a dislocation density between 10^{12} and 10^{13} /m^2 and after loading to 1 % plastic strain no clear increase in the dislocation density can be observed but slight changes in the patterns between both images are visible. After 5 % plastic strain individual structures (yellow) with higher dislocation densities of roughly 10^{14}/m^2 developed within the grains. These cell structures are accompanied by glide steps on the sample surface as can be seen in Figure 6.4c. After further deformation to 12 % - 28 % plastic strain the number of these structures inside the grains increase with no change in misorientation within the individual structures. In general the density of GNDs increases with increasing plastic strain but the maximum density of ~10^{14} /m^2 is not exceeded.

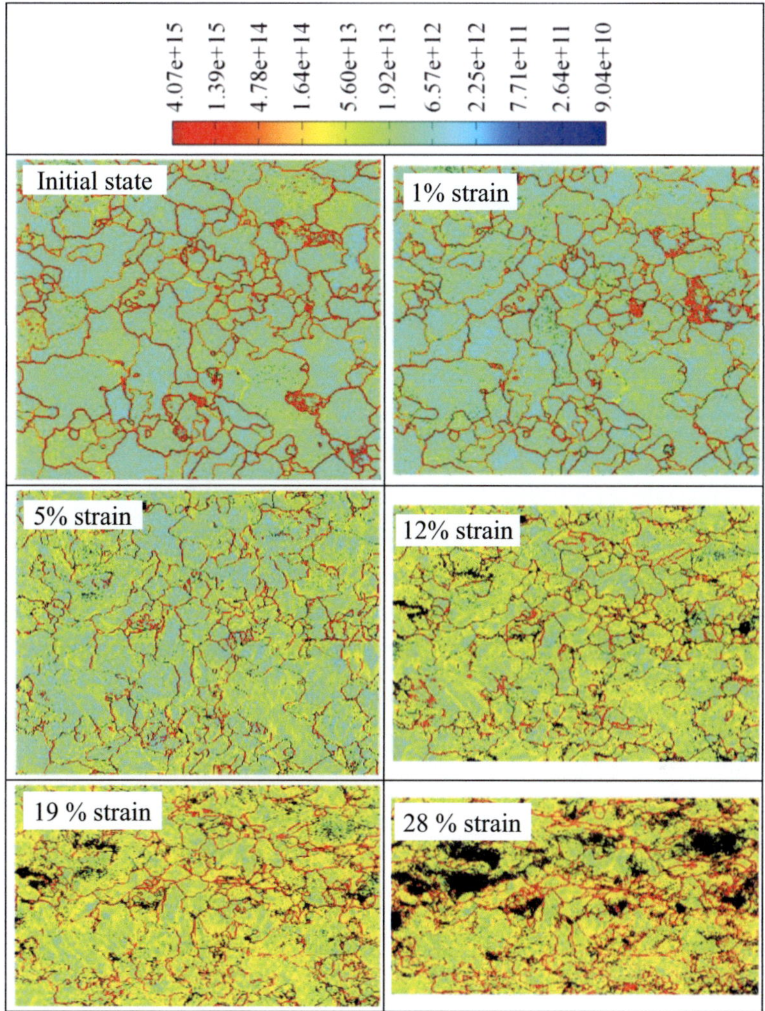

Figure 6.11: Calculated geometrically necessary dislocations from the presented EBSD data the previous images. The GNDs were calculated using equation (2.5).

6.3 Discussion

6.3.1 Texture evolution during tensile experiments

The evolution of the microstructure during the tensile experiment was investigated on a sample that was loaded parallel to the rolling direction (Figure 6.3). The microstructure was monitored by EBSD (Figure 6.5). The symmetry of the loading condition suggests that a texture along the tensile direction should form. It is expected that during plastic deformation the crystals rotate until several slip systems can be simultaneously active so that crystal rotation during deformation will be inhibited by the cooperative activation of several symmetrically equivalent slip systems. Deformation by rolling and deformation by tension testing both lead to the elongation of the sample along one axis. The deformation symmetry of both processes along this axis is the same, and therefore similarities in the texture exist. Rolling and tension have a component that leads to the elongation of the sample along one direction but rolling additionally contains a component that leads to the thinning of the sample on a plane perpendicular to the axis of elongation. From the point of view of symmetry, the rolling texture therefore may be considered as a special case of the texture that forms during deformation under tension. Due to the additional planar deformation its symmetry is expected to be reduced. The α-fiber is the texture component that is associated with elongation and both deformation textures show this component. The γ-fiber is a result of the compression that occur during rolling. In particular the interaction of both deformation directions leads to low symmetry and a detailed structure in the fiber plots in the rolling texture. Little or no structure is expected to form during tensile deformation due to the rotational symmetry of this deformation mode. For example no texture component will form relative to the plane of the sheet. The observation that the existing density of the <111> out of plane orientation decreases by ~12 % (γ-fiber) is in line with this consideration since it is expected that this texture component should decay due to the above symmetry reasons. Also the strong build-up of the α-fiber fits these considerations. In the observations the density of <110> orientations along the tensile axis increases by ~20%. This effect becomes evident when considering area fractions in the EBSD maps in figure 6.7 where the area fraction of green grains strongly increases during straining.

Similar conclusions can be drawn from the evolution of the α- and the γ-fibers (Figure 6.9). During tensile loading, the α-fiber in Figure 6.9 is shifted to higher intensities which implies that the grains rotate towards the <011> orientations during deformation. In the initial state, the α-fiber shows a maximum for the {111} <110> components which is still present after 28 % plastic strain, although the distribution becomes wider. This is in agreement with the above symmetry considerations where a <110> fiber texture is expected to form along the tensile direction. Such a texture would not have preferred orientations along the sheet normal and therefore a smooth distribution of densities in the α-fiber is expected. The decrease of the γ-fiber also fits to this scenario. Despite the overall decrease, the density around {111} <110> increases which suggests that some of the {111} grains start to rotate towards these symmetrically preferred orientations.

When comparing the deformation textures between cold rolling (Figure 4.5) and 28% of tensile strain (Figure 6.9) the expected differences in intensities (smaller intensity in γ-fibers and less narrow α-fiber) are not clearly visible. This may be due to the history of the sample. The sample tested under tension had been cold rolled and heat treated before the tensile test so that the assumption that the test started with a sample of random texture cannot be made here.

6.3.2 Anisotropic deformation behavior of DC04 steel

The anisotropic deformation behavior of the heat treated DC04 steel was investigated by tensile samples that were aligned parallel, 45° and 90° to the rolling directions. The results in Figure 6.1 show that the material has a higher yield and tensile strength when the loading direction is applied at an angle of 45 degree towards the rolling direction. Experiments on a second sample yield slightly different strengths that are ~50MPa higher (Figure 6.3). This difference may be due to the slightly different chemical composition of both samples.

The anisotropic deformation behavior of the sheet metal may be understood in terms of the behavior of the individual grains in this material. In the following, an attempt will be made to describe the stress strain response of the bulk material based on the mechanical behavior of the single crystalline pillars from chapter 5. This rather crude estimation is based on data from micro compression experiments of single crystalline pillars with different orientations and on the microstructure and in particular the volume fractions of the individual orientations in the sample. The size dependent deformation behavior of the DC04 steel was observed up to pillars diameter of 2 µm (Figure 5.5), for pillars larger than 2 µm no further decrease in strength was observed. Therefore it may be assumed, that the stress-strain data of pillars with diameters of 4 µm (Figure 5.2) represent the bulk behavior. The data for <111> and <123> orientations was presented and discussed in chapter 5. Unfortunately for pillars with <001> orientation only one sample was tested by microcompression and some uncertainty remains for this orientation. Also the stress-strain response of samples with <011> orientations was only investigated in pillars with a diameter of 2 µm. Altogether, no ideal orientation dependent data is available but nevertheless it will be attempted to explain the overall polycrystalline behavior based on measurements performed on the single crystalline micro pillars. The data that were used for this model are given in

Figure A1.1. The stress strain response (Figure 5.2c,d) and the deformation morphology (Figure 5.3g,h and Figure 5.4a) of polycrystalline micro pillars showed that grain boundaries do not strongly effect the strengths and deformation behavior of the samples. Therefore, the interaction between individual grains is neglected in this estimation.

Besides stress-strain data from single crystals of different orientations, microstructural information is needed. Since volume fractions cannot be easily obtained from the EBSD maps without making further assumptions, area fractions are instead used to estimate the volume fractions. The texture of the heat treated DC04 steel was investigated by EBSD measurements on an area of roughly 1 mm^2 (Figure 4.1). The EBSD results that are shown in Figure 4.1c are used to analyze the area fractions of the <111>, <110>, <001> and <123> orientations parallel, 45° and 90° to the rolling direction. All orientations that lie within 10° relative to the corresponding orientations were taken into account. It should be noted that an angle of 10° is rather large for bcc metals where the slip systems lie rather close but in order to capture enough orientations this assumption had to be made.

	Parallel to the RD	45° to the RD	90° to the RD
	Area fraction (%)		
<111>	0.79	0.55	5.32
<110>	19.12	15.63	13.20
<001>	5.43	0.85	1.87
<123>	20.68	22.81	12.89

Table 6.2: Area fraction of grains with orientations that are less than 10° away from <111>, <110>, <001> and <123>. Data was determined from Figure 4.1

In this estimation, the macroscopic stress strain data are calculated by assuming that all grains experience the same strain during deformation (Voigt model). The averaged stress-strain data for each orientation are multiplied by the area fraction $A_{<xyz>}$ of the corresponding orientation.

The resulting macroscopic stresses $\sigma_{macro}\,(\varepsilon)$ were obtained by a linear combination of all components according to

$$\sigma_{macro}\,(\varepsilon) = \frac{\sigma_{<111>}(\varepsilon) \cdot A_{<111>} + \sigma_{<123>}(\varepsilon) \cdot A_{<123>} + \sigma_{<011>}(\varepsilon) + A_{<011>} + \sigma_{<001>}(\varepsilon) \cdot A_{<001>}}{A_{<111>} + A_{<123>} + A_{<011>} + A_{<001>}}. \qquad (6.1)$$

In order to account for the effect that only four directions are considered and not all orientations data points are used (i.e. the columns in Table 6.2 do not add up to 100%) a normalization factor is used in equation (6.1). This estimation leads to the stress-strain responses that are plotted in Figure 6.12 which contains the result of this superposition for different sample directions. The results show that the estimated stress strain response differs from the macroscopic experimental results. The elastic modulus and the yield stress of the calculated data are distinctly lower than what was observed in the tensile experiments. This was already observed in the micro compression experiments in Figure 5.4b in which the deformation behavior of polycrystals during micro compression was compared to the tensile experiments. Lower elastic moduli and lower yield strengths are generally observed in micro compression experiments [54, 101]. These effects are commonly ascribed to the interaction between flat punch tip and pillar top or to the concurrent compression of material underneath the pillar. A further cause for such effects may be due to slight misalignments between pillar and tip. All these effects are not taken into account in the microcompression data and may lead to characteristic changes as observed in the stress strain curves that were used here.

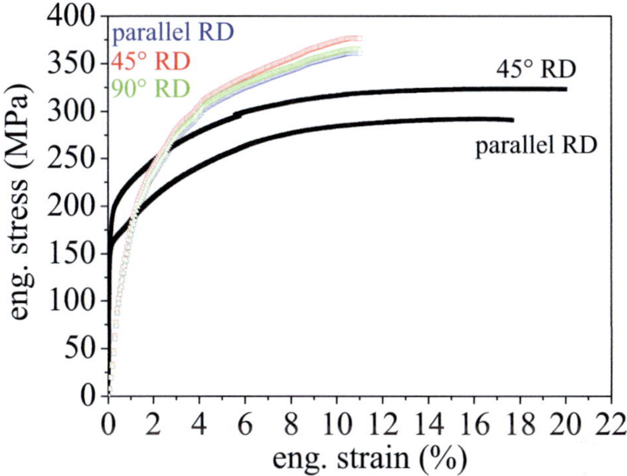

Figure 6.12: Stress-strain data (Figure 6.3) that were obtained from tensile experiments parallel and 45° relative to the rolling direction (solid lines) and simulated stress strain data from microcompression experiments on single crystalline pillars (red and green symbols)

The estimated tensile strength for tests at 45° relative to the rolling direction is ~16 MPa higher than what was found in the other directions. This difference is due to the different area fractions of the <111>, <123>, <011> and <001> orientations along the different sample directions. It should be noted that this estimation strongly depends on the input data from the microcompression experiment. The reference data that was used here is data from chapter 5. It was not recorded for this purpose and therefore no exact angle dependence is available and a rather large area in orientation space (10°) was associated to each orientation. It is therefore quite surprising that the higher strength for the test at 45° could be reproduced with this crude estimation. The fact that a simple linear combination of the single crystal deformation data exhibits the trends in the sample anisotropy that are found in the bulk materials shows that the assumption may be justified that grain interactions may not be important in this system. From this estimation it may be concluded that the anisotropic behavior is mostly determined by the texture of the sample and that grain shapes and grain interactions may not play a strong role in the deformation behavior of DC04 sheets. Also surprising is the fact that strengths are obtained that are not very far from the strengths of DC04 sheet metal. This further supports the observation that size dependent behavior does not exist for samples with dimensions larger than 4µm (Chapter 5).

It is surprising that this simple concept produces reasonable stress levels and shows anisotropy trends as observed in the tensile tests. Such concepts have been successfully used before. For example Hansen et al. [102] used a similar concept to determine the macroscopic stress-strain curve of a polycrystalline aluminum sample based on tensile data of single crystalline samples which were weighted with the volume fractions of different orientations. The calculated data predicted the macroscopic stress-strain response up to strains of 10% fairly well but overestimated the flow stresses at larger strains. In order to more accurately simulate the anisotropic deformation behavior, more complex estimations have to be made that consider the activation of the different glide systems and grain shapes using crystal plasticity modeling. This thesis is part of Graduiertenkolleg 1483 of the Deutsche Forschungsgemeinschaft and within this research training group such micro mechanical models are developed in project A2 [103, 104].

7. **Summary**

Ferritic deep drawing steels are commonly used in the automotive industry. Their deformation behavior is strongly influenced by their microstructure which is a result of alloying elements and mechanical and thermal treatments. In order to manufacture a sheet metal with desired mechanical properties, the influence of the different microstructural components on the deformation behavior of the sheet metal needs to be understood and the microstructure needs to be controlled. The aim of this work was to supply data in order to model groups for simulating steel production but also to gain a better fundamental understanding of how microstructure and mechanical behavior are related. Along this path of research, additional insights into the fundamentals of deformation and microstructural evolution of DC04 steel were obtained.

The microstructural characterization showed that the process conditions have a very strong impact on the microstructure of the material. The hot rolled material consists of small grains with no preferred crystal orientations. After the cold rolling process the grains are elongated along the rolling direction and contain large misorientations corresponding to high dislocation densities ($\geq 10^{14}/m^2$). As expected from cold rolling, pronounced α-fibers and γ-fibers were present. After the industrial heat treatment process, the microstructure consists of large grains and exhibits a stronger γ-fiber and a less distinctive α-fiber than the cold rolled state. The evolution of the microstructure during heat treatment was investigated in detail by heating experiments at two different temperatures (647°C and 680°C) and at different heating rates. These experiments gave insight into the development of different texture components and into local processes such as grain nucleation and growth. During the slow heating experiments at 647°C two distinctly different effects were observed. The evolution started with a small increase of density in the $\{111\}$ <110> orientations in the α-fiber and an isotropic increase of the γ-fiber. This was accompanied by an intense growth of the $\{111\}$ grains in the EBSD maps. These observations indicate that strongly deformed regions with $\{111\}$ orientation start to recover and grow in excess of their neighboring grains. The nucleation of subgrains or reorientations of grains was rarely observed in the microscopic observations. After further heating only minor changes were found in the shape and spatial distribution of the grains but strong changes in the crystal orientation occurred where the density of $\{111\}$ <110> orientations strongly increased. Surprisingly, the dislocation density in the grains kept reducing even after the grains already had stopped growing. Altogether the observations show that at 647°C the formation of the new macrostructure is controlled by grain growth and subsequent reorientations of grains. After the fast heating experiment to 680°C, a microstructure with smaller grains had developed. The isotropic decrease of density in the α-fiber and γ-fiber showed that the microstructure consisted of grains with almost random orientations. The observation of the two very different microstructures led to the assumption that the dominant mechanisms may change when the samples are heated to the two different temperatures. Therefore, two possible effects were discussed that may explain the changes in the observed microstructures. First, classic arguments were considered where recovery at low temperatures reduces defects and therefore the number of nucleation sites for recrystallization. Slow heating at low temperatures leads to only few nucleation sites and large grains will develop later during grain growth. In the fast heating to the higher temperature, there is little time for recovery to take place and more competing nuclei are present so that only small grains will form during recrystallization. These arguments can explain the observed microstructures but since classical nucleation in the form of subgrain formation was not found in the samples, an alternative description was elaborated and used in a simple Gedankenexperiment. Here the microstructure evolution is governed by the competition between grain growth and recovery. At the lower heating temperature, grain growth is the dominant mechanism which results in large grains whereas at higher temperatures, dislocation mobility becomes dominant which leads to recovery without grain growth. The competition between both mechanisms leads to larger grains at 647°C and grains with sizes similar to those of the starting material at 680°C, consistent with the experimental observations.

The size dependent deformation behavior of the heat treated DC04 steel was investigated by microcompression tests on pillars with different size. Experiments on single and polycrystalline pillars revealed the effect of alloying elements, precipitates, grain boundaries and stored plastic energy on the size dependent deformation behavior DC04 steel. The results were compared to the size dependent deformation behavior of pure α-iron. In the heat treated DC04 steel, pillars with diameters larger than 2 μm show strengths that are higher than what was found for α-Fe. These strengths are in agreement with the bulk strength as measured by tensile experiments on macroscopic DC04 samples. The increase in strength compared to pure iron is most likely caused by the solid solution of carbon but it may also be caused by particle strengthening of small MnS precipitates that were found in the matrix. For pillars with diameters smaller than 2 μm, a size dependent deformation behavior similar to α-Fe was found. Strengths and scaling exponents depended on the orientation of the pillars. Pillars with <111> and <123> orientations exhibited a scaling similar to <211> orientatied α-Fe with with an exponent around -0.8. DC04 pillars with <001> orientations have lower strengths and a smaller scaling exponent. The different strengths were explained by different resolved shear stresses assuming that glide occurs on {011} and/or {112} planes. The different scaling may be explained by the fact that more nucleation sites exist in high symmetry orientations than in pillars oriented for single slip. If the high stresses are due to the fact that dislocation nucleation is the strength controlling mechanism, concepts such dislocation starvation can easily explain the lower stresses found in the small samples with high symmetry. The deviation of the strength from bulk strength at the onset of the size dependent behavior also seems to dependent on the orientation and occurs at the smallest size for <100> oriented pillars. The deformation morphologies were investigated for pillars with diameters of 2 μm, 4 μm and 22 μm. From the morphologies it was observed that grain boundaries are no obstacles for dislocation motion since glide processes can run through grain boundaries. This interpretation is in agreement with the fact that no strengthening was measured when grain boundaries were present in the pillars.

The effect of residual defects was investigated by experiments on cold rolled D04 steel pillars. For pillars with a diameter of 4 μm the strengths are higher than what was determined for the heat treated material. This increase is explained by work hardening and the formation of a cell structure in the highly deformed material. In the cold rolled steel, almost no size dependence of the strength was found. Small pillars of the heat treated material are stronger than the cold rolled material. This is in agreement with the concept of dislocation starvation since dislocation nucleation may not be a limiting factor for small pillars that already contain a high number of dislocations.

The macroscopic deformation behavior was investigated by tensile experiments that were performed along the rolling direction. The evolution of the microstructure was monitored by EBSD after five unloading steps. During deformation, the α-fiber strengthened according to what is expected from symmetry considerations. The existing density in the γ-fiber decreased with increasing plastic strain. This is expected since there is no symmetry in tensile straining relative to the sheet normal. The tensile experiments that were performed along 0°, 45° and 90° to the rolling direction showed that heat treated material exhibits the highest strength when the sample is loaded at 45° to the rolling direction. In a simple model, the microcompression data of single crystalline pillars with different orientations were combined with the corresponding area fractions of the sheet metals using a Voigt model. This rough estimation shows that the mechanical anisotropy is mainly determined by the texture components of the sheet metal.

In this work, site specific pillar testing was used to determine the mechanical properties of selected microstructural elements. Based on the local mechanical data it was attempted to describe the macroscopic deformation behavior of a material. For the rather simple low alloy steel considered in this thesis this concept could be used and even a simple Voigt model could reproduce trends in the plastic anisotropy of the bulk material. In order to more accurately predict mechanical properties or to apply this concept to advanced materials, more elaborate models and simulations are needed. Within Graduiertenkolleg 1483 attempts are made to establish simulations for more advanced steels. Near term goals are to extend this concept to duplex (ferrite/austenite) or dual phase (ferrite/martensite) steels and as long term goals complex multiphase materials such as TRIP (transformation induced plasticity) or TWIP (twinning induced plasticity) steels may even be considered. If such complex materials can be simulated based on local mechanical data an important goal comes into reach which is the use of simulations to predict suitable production routes for manufacturing steels with tailored mechanical properties.

8.   Bibliography

[1]   P.J. Jacques, *Transformation-induced plasticity for high strength formable steels.* Current Opinion in Solid State & Materials Science, 2004. 8(3-4): p. 259-265.

[2]   Q. Furnemont, M. Kempf, P.J. Jacques, M. Goken, and F. Delannay, *On the measurement of the nanohardness of the constitutive phases of TRIP-assisted multiphase steels.* Materials Science and Engineering a-Structural Materials Properties Microstructure and Processing, 2002. 328(1-2): p. 26-32.

[3]   P.J. Jacques, Q. Furnemont, F. Lani, T. Pardoen, and F. Delannay, *On the measurement of the nanohardness of the constitutive phases of TRIP-assisted multiphase steels.* Acta Materialia, 2007. 55(11): p. 3681-3693.

[4]   O. Grassel, L. Kruger, G. Frommeyer, and L.W. Meyer, *High strength Fe-Mn-(Al, Si) TRIP/TWIP steels development - properties - application.* International Journal of Plasticity, 2000. 16(10-11): p. 1391-1409.

[5]   D. Barbier, N. Gey, S. Allain, N. Bozzolo, and M. Humbert, *Analysis of the tensile behavior of a TWIP steel based on the texture and microstructure evolutions.* Materials Science and Engineering a-Structural Materials Properties Microstructure and Processing, 2009. 500(1-2): p. 196-206.

[6]   W. Bleck, B. Engl, A. Frehn, D. Nicklas, and G. Steinbeck, Ermittlung von Berechnungskennwerten an Karosseriewerkstoffen – Bericht über ein Gemeinschaftsprojekt der Stahl- und Automobilindustrie. Materialwissenschaft und Werkstofftechnik. 35(8): p. 483–494,.

[7]   M.D. Uchic, D.M. Dimiduk, J.N. Florando, and W.D. Nix, *Sample dimensions influence strength and crystal plasticity.* Science, 2004. 305(5686): p. 986-989.

[8]   *Electron Backscatter Diffraction in Materials Science* A.J. Schwartz, M. Kumar, B.L. Adams, and D.P. Field, Editors. 2009, Springer.

[9]   O. Engler and V. Randle, Introduction to Texture Analysis: Macrotexture, Microtexture, and Orientation Mapping. 2009, CRC Press Taylor & Francis Group: New York.

[10]   G. Gottstein, Physikalische Grundlagen der Materialkunde, in Springer. 2001, Springer. p. 37.

[11]   H.J. Bunge, ed. Texture Analysis in Materials Science: Mathematical Methods. 1993, CUVILLIER VERLAG: Gottingen.

[12]   HKLTechnology, *CHANNEL 5.* 2004.

[13]   H.J. Bunge, *PARTIAL TEXTURE ANALYSIS.* Textures and Microstructures, 1990. 12(1-3): p. 47-63.

[14]   O. Engler and V. Randle, Evaluation and Representation of Macrotexture Data, in Introduction to Texture Analysis: Macrotexture, Microtexture, and Orientation Mapping. 2009, CRC Press Taylor & Francis Group: New York. p. 127.

[15]   J.F. Nye, *SOME GEOMETRICAL RELATIONS IN DISLOCATED CRYSTALS.* Acta Metallurgica, 1953. 1(2): p. 153-162.

[16]   E. Demir, D. Raabe, N. Zaafarani, and S. Zaefferer, Investigation of the indentation size effect through the measurement of the geometrically necessary dislocations beneath small indents of different depths using EBSD tomography. Acta Materialia, 2009. 57(2): p. 559-569.

[17]   M. Calcagnotto, D. Ponge, E. Demir, and D. Raabe, Orientation gradients and geometrically necessary dislocations in ultrafine grained dual-phase steels studied by 2D and 3D EBSD. Materials Science and Engineering a-Structural Materials Properties Microstructure and Processing, 2010. 527(10-11): p. 2738-2746.

[18]   W. Pantleon, Resolving the geometrically necessary dislocation content by conventional electron backscattering diffraction. Scripta Materialia, 2008. 58(11): p. 994-997.

[19]   W. He, W. Ma, and W. Pantleon, Microstructure of individual grains in cold-rolled aluminium from orientation inhomogeneities resolved by electron backscattering diffraction. Materials Science and Engineering a-Structural Materials Properties Microstructure and Processing, 2008. 494(1-2): p. 21-27.

[20]   N. Hansen and D.J. Jensen, *FLOW-STRESS ANISOTROPY CAUSED BY GEOMETRICALLY NECESSARY BOUNDARIES.* Acta Metallurgica Et Materialia, 1992. 40(12): p. 3265-3275.

[21]   O. Engler and V. Randle, Introduction to Texture Analysis: Macrotexture, Microtexture, and Orientation Mapping. 2009, CRC Press Taylor & Francis Group: New York. p. 156.

[22]   M. Holscher, D. Raabe, and K. Lucke, *ROLLING AND RECRYSTALLIZATION TEXTURES OF BCC STEELS.* Steel Research, 1991. 62(12): p. 567-575.

[23]   D. Raabe and K. Lucke, ROLLING AND ANNEALING TEXTURES OF BCC METALS, in Proceedings of the 10th International Conference on Textures of Materials, Pts 1 and 2 - Icotom-10, H.J. Bunge, Editor. 1994, Transtec Publications Ltd: Zurich-Uetikon. p. 597-610.

[24] D. Raabe and K. Lucke, *TEXTURE AND MICROSTRUCTURE OF HOT ROLLED STEEL.* Scripta Metallurgica Et Materialia, 1992. 26(8): p. 1221-1226.

[25] G. Gottstein, Physikalische Grundlagen der Materialkunde, in Springer. 2001, Springer. p. 313.

[26] A. Haldar, S. Suwas, and D. Bhattacharjee, *Proceedings of the International Conference on Microstructure and Texture in Steels and Other Materials*, ed. A.S. Haldar, Satyam; Bhattacharjee, Debashish. 2008, Jameshedpur, India: Springer.

[27] U. Von Schlippenbach, F. Emren, and K. Lucke, INVESTIGATION OF THE DEVELOPMENT OF THE COLD-ROLLING TEXTURE IN DEEP DRAWING STEELS BY ODF-ANALYSIS. Acta Metallurgica, 1986. 34(7): p. 1289-1301.

[28] M.H. F.J. HUMPHREYS, *Recrystallization and Related Annealing Phenomena*, ed. Edition. 2004: Elsevier.

[29] L.B. Freund and S. Suresh, *Thin Film Materials: Stress, Defect Formation, and Surface Evolution*. 2008, Cambridge University Press. p. 48.

[30] Gottstein, Rekristallisation metallischer Werkstoffe: Grundlagen, Analyse, Anwendung. 1984: Dt. Ges. für Metallkunde.

[31] Gottstein, in Physikalische Grundlagen der Materialkunde. 2001, Springer. p. 289ff.

[32] M. Wenk, Mikrostrukturentwicklung bei der Wärmebehandlung von Stahl, in IAM-WBM. 2012, Karlsruhe Institute of Technology (KIT).

[33] T.H. Courtney, *Mechanical Behavior of Materials*. 2000, Waveland Press, Inc. p. 186-196.

[34] R.L. Fleischer, *SUBSTITUTIONAL SOLUTION HARDENING.* Acta Metallurgica, 1963. 11(3): p. 203-&.

[35] J.O. Linde and S. Edwardson, INVESTIGATION OF THE CRITICAL SHEAR STRESS FOR SINGLE CRYSTALS OF METALLIC SOLID SOLUTIONS .2. Arkiv for Fysik, 1954. 8(6): p. 511-519.

[36] C.A. Wert, *SOLID SOLUBILITY OF CEMENTITE IN ALPHA-IRON.* Transactions of the American Institute of Mining and Metallurgical Engineers, 1950. 188(10): p. 1242-1244.

[37] E. Arzt, Overview no. 130 - Size effects in materials due to microstructural and dimensional constraints: A comparative review. Acta Materialia, 1998. 46(16): p. 5611-5626.

[38] B.L. Li, A. Godfrey, Q.C. Meng, Q. Liu, and N. Hansen, *Microstructural evolution of IF-steel during cold rolling.* Acta Materialia, 2004. 52(4): p. 1069-1081.

[39] D. Kuhlmann-Wilsdorf, *The theory of dislocation-based crystal plasticity.* Philosophical Magazine a-Physics of Condensed Matter Structure Defects and Mechanical Properties, 1999. 79(4): p. 955-1008.

[40] J.P. Hirth and J. Lothe, *Theory of dislocations.* 1982, McGraw-Hill. p. 788-789.

[41] H.W. Song, S.R. Guo, and Z.Q. Hu, A coherent polycrystal model for the inverse Hall-Petch relation in nanocrystalline materials. Nanostructured Materials, 1999. 11(2): p. 203-210.

[42] P.A. Gruber, J. Bohm, F. Onuseit, A. Wanner, R. Spolenak, and E. Arzt, Size effects on yield strength and strain hardening for ultra-thin Cu films with and without passivation: A study by synchrotron

and bulge test techniques. Acta Materialia, 2008. 56(10): p. 2318-2335.

[43]   G. Dehm, T.J. Balk, H. Edongue, and E. Arzt, *Small-scale plasticity in thin Cu and Al films.* Microelectronic Engineering, 2003. 70(2-4): p. 412-424.

[44]   W.D. Nix, *MECHANICAL-PROPERTIES OF THIN-FILMS.* Metallurgical Transactions a-Physical Metallurgy and Materials Science, 1989. 20(11): p. 2217-2245.

[45]   D. Kiener, C. Motz, T. Schoberl, M. Jenko, and G. Dehm, *Determination of mechanical properties of copper at the micron scale.* Advanced Engineering Materials, 2006. 8(11): p. 1119-1125.

[46]   W.D. Nix, J.R. Greer, G. Feng, and E.T. Lilleodden, Deformation at the nanometer and micrometer length scales: Effects of strain gradients and dislocation starvation. Thin Solid Films, 2007. 515(6): p. 3152-3157.

[47]   J.R. Greer and W.D. Nix, Nanoscale gold pillars strengthened through dislocation starvation. Physical Review B, 2006. 73(24).

[48]   J.R. Greer and W.D. Nix, Size dependence in mechanical properties of gold at the micron scale in the absence of strain gradients. Applied Physics a-Materials Science & Processing, 2008. 90(1): p. 203-203.

[49]   C.A. Volkert and E.T. Lilleodden, *Size effects in the deformation of sub-micron Au columns.* Philosophical Magazine, 2006. 86(33-35): p. 5567-5579.

[50]   B.E. Schuster, Q. Wei, H. Zhang, and K.T. Ramesh, *Microcompression of nanocrystalline nickel.* Applied Physics Letters, 2006. 88(10).

[51]   D.M. Dimiduk, M.D. Uchic, and T.A. Parthasarathy, *Size-affected single-slip behavior of pure nickel microcrystals.* Acta Materialia, 2005. 53(15): p. 4065-4077.

[52]   R. Maass and M.D. Uchic, In-situ characterization of the dislocation-structure evolution in Ni micro-pillars. Acta Materialia, 2012. 60(3): p. 1027-1037.

[53]   K.S. Ng and A.H.W. Ngan, *Stochastic nature of plasticity of aluminum micro-pillars.* Acta Materialia, 2008. 56(8): p. 1712-1720.

[54]   M.D. Uchic, P.A. Shade, and D.M. Dimiduk, Plasticity of Micrometer-Scale Single Crystals in Compression, in Annual Review of Materials Research. 2009, Annual Reviews: Palo Alto. p. 361-386.

[55]   M.D. Uchic, P.A. Shade, and D.M. Dimiduk, Micro-compression testing of fcc metals: A selected overview of experiments and simulations. Jom, 2009. 61(3): p. 36-41.

[56]   D. Kaufmann, Size Effects on the Plastic Deformation of the BCC-Metals Ta and Fe. 2011, Cuvillier, E: Karlsruhe.

[57]   O. Kraft, P.A. Gruber, R. Monig, and D. Weygand, *Plasticity in Confined Dimensions*, in *Annual Review of Materials Research, Vol 40*, D.R. Clarke, M. Ruhle, and F. Zok, Editors. 2010, Annual Reviews: Palo Alto. p. 293-317.

[58]   D. Kaufmann, R. Monig, C.A. Volkert, and O. Kraft, *Size dependent mechanical behaviour of tantalum.* International Journal of Plasticity, 2011. 27(3): p. 470-478.

[59]   A.S. Schneider, D. Kaufmann, B.G. Clark, C.P. Frick, P.A. Gruber, R. Monig, O. Kraft, and E. Arzt, *Correlation between critical*

temperature and strength of small-scale bcc pillars. Physical Review Letters, 2009. 103(10): p. 105501.

[60] A.S. Schneider, C.P. Frick, B.G. Clark, P.A. Gruber, and E. Arzt, *Influence of orientation on the size effect in bcc pillars with different critical temperatures.* Materials Science and Engineering a-Structural Materials Properties Microstructure and Processing, 2011. 528(3): p. 1540-1547.

[61] S. Brinckmann, J.Y. Kim, and J.R. Greer, Fundamental differences in mechanical behavior between two types of crystals at the nanoscale. Physical Review Letters, 2008. 100(15).

[62] J.Y. Kim, D.C. Jang, and J.R. Greer, Insight into the deformation behavior of niobium single crystals under uniaxial compression and tension at the nanoscale. Scripta Materialia, 2009. 61(3): p. 300-303.

[63] M. Tang, L.P. Kubin, and G.R. Canova, Dislocation mobility and the mechanical response of BCC single crystals: A mesoscopic approach. Acta Materialia, 1998. 46(9): p. 3221-3235.

[64] W.A. Spitzig and A.S. Keh, EFFECT OF ORIENTATION AND TEMPERATURE ON PLASTIC FLOW PROPERTIES OF IRON SINGLE CRYSTALS. Acta Metallurgica, 1970. 18(6): p. 611-&.

[65] E. Kuramoto, Y. Aono, K. Kitajima, K. Maeda, and S. Takeuchi, *THERMALLY ACTIVATED SLIP DEFORMATION BETWEEN 0.7-K AND 77-K IN HIGH-PURITY IRON SINGLE-CRYSTALS.* Philosophical Magazine a-Physics of Condensed Matter Structure Defects and Mechanical Properties, 1979. 39(6): p. 717-724.

[66] C. Domain and G. Monnet, Simulation of screw dislocation motion in iron by molecular dynamics simulations. Physical Review Letters, 2005. 95(21).

[67] D. Caillard, An in situ study of hardening and softening of iron by carbon interstitials. Acta Materialia, 2011. 59(12): p. 4974-4989.

[68] J.W. Christian, *SOME SURPRISING FEATURES OF THE PLASTIC-DEFORMATION OF BODY-CENTERED CUBIC METALS AND ALLOYS*. Metallurgical Transactions a-Physical Metallurgy and Materials Science, 1983. 14(7): p. 1237-1256.

[69] D. Caillard, Kinetics of dislocations in pure Fe. Part I. In situ straining experiments at room temperature. Acta Materialia, 2010. 58(9): p. 3493-3503.

[70] D. Brunner and V. Glebovsky, Analysis of flow-stress measurements of high-purity tungsten single crystals. Materials Letters, 2000. 44(3-4): p. 144-152.

[71] L. Hollang, M. Hommel, and A. Seeger, *The flow stress of ultra-high-purity molybdenum single crystals.* Physica Status Solidi a-Applied Research, 1997. 160(2): p. 329-354.

[72] H. Bei, S. Shim, G.M. Pharr, and E.P. George, Effects of pre-strain on the compressive stress-strain response of Mo-alloy single-crystal micropillars. Acta Materialia, 2008. 56(17): p. 4762-4770.

[73] T.A. Parthasarathy, S.I. Rao, D.M. Dimiduk, M.D. Uchic, and D.R. Trinkle, *Contribution to size effect of yield strength from the stochastics of dislocation source lengths in finite samples.* Scripta Materialia, 2007. 56(4): p. 313-316.

[74] J. Senger, D. Weygand, P. Gumbsch, and O. Kraft, *Discrete dislocation simulations of the plasticity of micro-pillars under uniaxial loading.* Scripta Materialia, 2008. 58(7): p. 587-590.

[75] J. Senger, D. Weygand, C. Motz, P. Gumbsch, and O. Kraft, *Aspect ratio and stochastic effects in the plasticity of uniformly loaded*

micrometer-sized specimens. Acta Materialia, 2011. 59(8): p. 2937-2947.

[76] M. Wegst and C.W. Wegst, *Stahlschlüssel.* 2007.

[77] J.F. Ziegler, J.P. Biersack, and U. Littmark, *The Stopping Range of Ions in Matter*, in 1985, Pergamon Press: New York.

[78] J.P. McCaffrey, M.W. Phaneuf, and L.D. Madsen, Surface damage formation during ion-beam thinning of samples for transmission electron microscopy. Ultramicroscopy, 2001. 87(3): p. 97-104.

[79] D. Kiener, C. Motz, M. Rester, M. Jenko, and G. Dehm, *FIB damage of Cu and possible consequences for miniaturized mechanical tests.* Materials Science and Engineering a-Structural Materials Properties Microstructure and Processing, 2007. 459(1-2): p. 262-272.

[80] S. Shim, H. Bei, M.K. Miller, G.M. Pharr, and E.P. George, Effects of focused ion beam milling on the compressive behavior of directionally solidified micropillars and the nanoindentation response of an electropolished surface. Acta Materialia, 2009. 57(2): p. 503-510.

[81] C. Motz, T. Schoberl, and R. Pippan, Mechanical properties of micro-sized copper bending beams machined by the focused ion beam technique. Acta Materialia, 2005. 53(15): p. 4269-4279.

[82] D. Hall and D.J. Bacon, Dislocation Arrays and Crystal Boundaries, in Introduction to Dislocatoins. 2001, Butterworth-Heinemann. p. 160.

[83] W.C. Oliver and G.M. Pharr, An Improved Technique for Determining Hardness and Elastic-Modulus Using Load and Displacement Sensing Indentation Experiments. Journal of Materials Research, 1992. 7(6): p. 1564-1583.

[84] R. Schwaiger, M. Weber, B. Moser, P. Gumbsch, and O. Kraft, *Mechanical assessment of ultrafine-grained nickel by microcompression experiment and finite element simulation.* Journal of Materials Research, 2012. 27(1): p. 266-277.

[85] J.-Y. Kang, D.-I. Kim, and H.-C. Lee, *Development of Recrytsallization Texture,* in *Recrystallization and Related Annealing Phenomena,* F.J. Humphreys and M. Hatherly, Editors. 2004. p. 93.

[86] H. Nakamichi, F.J. Humphreys, P.S. Bate, and I. Brough, *In-situ EBSD observation of the recrystallization of an IF steel at high temperature* Materials Science Forum, 2007. 550: p. 441-446.

[87] Dillamor.Il, C.J.E. Smith, Hutchins.Wb, and P.L. Morris, *TRANSITION BANDS AND RECRYSTALLIZATION IN METALS.* Proceedings of the Royal Society of London Series a-Mathematical and Physical Sciences, 1972. 329(1579): p. 405-&.

[88] A. Inoue, H. Nitta, and Y. Iijima, *Grain boundary self-diffusion in high purity iron.* Acta Materialia, 2007. 55(17): p. 5910-5916.

[89] Y. Iijima, K. Kimura, and K. Hirano, *SELF-DIFFUSION AND ISOTOPE EFFECT IN ALPHA-IRON.* Acta Metallurgica, 1988. 36(10): p. 2811-2820.

[90] Y. Shima, Y. Ishikawa, H. Nitta, Y. Yamazaki, K. Mimura, M. Isshiki, and Y. Iijima, *Self-diffusion along dislocations in ultra high purity iron.* Materials Transactions, 2002. 43(2): p. 173-177.

[91] A.D. Rollett, G. Gottstein, L.S. Shvindlerman, and D.A. Molodov, *Grain boundary mobility - a brief review.* Zeitschrift Für Metallkunde, 2004. 95(4): p. 226-229.

[92]   D. Turnbull, *THEORY OF GRAIN BOUNDARY MIGRATION RATES*. Transactions of the American Institute of Mining and Metallurgical Engineers, 1951. 191(8): p. 661-665.

[93]   J.P. Hirth and J. Lothe, *Theory of dislocations*. 1982: McGraw-Hill. p. 561.

[94]   M. Merkel and K.-H. Thomas, *Taschenbuch der Werkstoffe*. 2000, Fachbuchverlag Leipzig im Carl Hanser Verlag. p. 158.

[95]   R.M. Keller, S.P. Baker, and E. Arzt, Quantitative analysis of strengthening mechanisms in thin Cu films: Effects of film thickness, grain size, and passivation. Journal of Materials Research, 1998. 13(5): p. 1307-1317.

[96]   R. Venkatraman and J.C. Bravman, SEPARATION OF FILM THICKNESS AND GRAIN-BOUNDARY STRENGTHENING EFFECTS IN AL THIN-FILMS ON SI. Journal of Materials Research, 1992. 7(8): p. 2040-2048.

[97]   B. Girault, A.S. Schneider, C.P. Frick, and E. Arzt, *Strength Effects in Micropillars of a Dispersion Strengthened Superalloy*. Advanced Engineering Materials, 2010. 12(5): p. 385-388.

[98]   J.R. Greer, W.C. Oliver, and W.D. Nix, Size dependence of mechanical properties of gold at the micron scale in the absence of strain gradients. Acta Materialia, 2005. 53(6): p. 1821-1830.

[99]   H. Gao, Y. Huang, W.D. Nix, and J.W. Hutchinson, *Mechanism-based strain gradient plasticity - I. Theory*. Journal of the Mechanics and Physics of Solids, 1999. 47(6): p. 1239-1263.

[100] S. Reyntjens and R. Puers, *A review of focused ion beam applications in microsystem technology*. Journal of Micromechanics and Microengineering, 2001. 11(4): p. 287-300.

[101] H. Zhang, B.E. Schuster, Q. Wei, and K.T. Ramesh, *The design of accurate micro-compression experiments.* Scripta Materialia, 2006. 54(2): p. 181-186.

[102] N. Hansen and X. Huang, *Microstructure and flow stress of polycrystals and single crystals.* Acta Materialia, 1998. 46(5): p. 1827-1836.

[103] T. Phan Van, K. öchen, and T. Böhlke. Validation of Material Models in Grain Scale Simulation based on EBSD Experimental Data. in PAMM. 2011: Proc. Appl. Math. Mech.

[104] T. Phan Van, K. Jöchen, and T. Böhlke. Micromechanical Modeling for Text Texture Evolution and Deformation Localization in Metal Forming Operations. in Summerschool Graduiertenschule 1483. 2011.

A   Appendix

A.1   Averaged stress strain data from micro compression experiments

Figure A1.1 shows stress strain data that were obtained by micro compression experiments on single crystalline pillars (chapter 5) with different orientations.

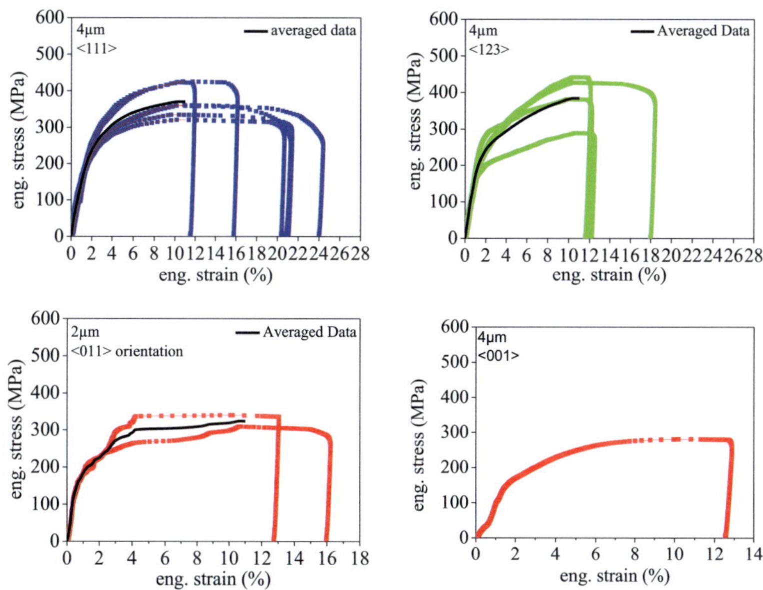

Figure A1.1: Stress strain data on single crystalline heat treated DC04 steel pillars and the averaged curves that are calculated for each orientation.

Samples with <111>, <123> and <001> orientations had diameter of 4 μm and pillars with <110> orientations have diameter of 2 μm. The stress-strain data were averaged for each orientation to obtain a representative curve. These averaged curves were used for the calculation of the macroscopic stress-strain data in section 6.3.2.

Schriftenreihe
des Instituts für Angewandte Materialien

ISSN 2192-9963

Die Bände sind unter www.ksp.kit.edu als PDF frei verfügbar oder
als Druckausgabe bestellbar.

Band 1   Prachai Norajitra
          Divertor Development for a Future Fusion Power Plant. 2011
          ISBN 978-3-86644-738-7

Band 2   Jürgen Prokop
          **Entwicklung von Spritzgießsonderverfahren zur Herstellung von
          Mikrobauteilen durch galvanische Replikation.** 2011
          ISBN 978-3-86644-755-4

Band 3   Theo Fett
          **New contributions to R-curves and bridging stresses – Applications
          of weight functions.** 2012
          ISBN 978-3-86644-836-0

Band 4   Jérôme Acker
          **Einfluss des Alkali/Niob-Verhältnisses und der Kupferdotierung auf
          das Sinterverhalten, die Strukturbildung und die Mikrostruktur von
          bleifreier Piezokeramik $(K_{0,5}Na_{0,5})NbO_3$.** 2012
          ISBN 978-3-86644-867-4

Band 5   Holger Schwaab
          **Nichtlineare Modellierung von Ferroelektrika unter Berücksich-
          tigung der elektrischen Leitfähigkeit.** 2012
          ISBN 978-3-86644-869-8

Band 6   Christian Dethloff
          **Modeling of Helium Bubble Nucleation and Growth in Neutron
          Irradiated RAFM Steels.** 2012
          ISBN 978-3-86644-901-5

Band 7   Jens Reiser
          **Duktilisierung von Wolfram. Synthese, Analyse und Charakterisierung
          von Wolframlaminaten aus Wolframfolie.** 2012
          ISBN 978-3-86644-902-2

Band 8   Andreas Sedlmayr
          Experimental Investigations of Deformation Pathways in Nanowires.
          2012
          ISBN 978-3-86644-905-3